FUNCTIONS OF
COMPLEX VARIABLES

By

Phillip FRANKLIN

Professor of Mathematics
Massachusetts Institute of Technology

Englewood Cliffs, N.J.
PRENTICE-HALL, INC.

©—1958, BY
PRENTICE-HALL, INC.
ENGLEWOOD CLIFFS, N.J.

ALL RIGHTS RESERVED. NO PART OF THIS BOOK
MAY BE REPRODUCED IN ANY FORM, BY MIMEO-
GRAPH OR ANY OTHER MEANS, WITHOUT PERMIS-
SION IN WRITING FROM THE PUBLISHERS

Library of Congress Catalog Card No.: 58-10381

2018 Preface, Recommended Reading,
Cover and About The Authors page @2018 Karo Maestro

PREFACE

The object of this book is to introduce the reader to the general properties of an analytic function of a single complex variable. The precise topics discussed, and the order chosen, may be seen from the table of contents. Because of applications of interest to engineers, physicists, and applied mathematicians there is considerable emphasis on such topics as conformal mapping and residue theory. Nevertheless, the definitions and deductions of theorems in sequence are given with sufficient care and precision to satisfy the student or teacher whose prime interest is in the theory as a branch of pure mathematics.

For each main topic the treatment is logical and essentially complete. But to avoid too extensive digression some theorems borrowed from the theory of functions of real variables have been stated and explained without including proof. In order to avoid undue complication, the theorems involving complex variables have been formulated as ordinarily applied, rather than with the utmost known generality. But an honest attempt has been made to preserve the rigor and elegance of the subject, and to avoid artificial approaches or non-motivated short cuts.

The book may be used as a class text for a first course in complex variables at the upper class or graduate level. Several discussions, such as that of complex numbers in chapter one, of power series in chapter three, of elementary functions in chapter four, and of integrals in chapter seven, are sufficiently detailed to be accessible to readers with no mathematical background other than a first course

PREFACE

in the calculus. However, for students who have already had a brief formal treatment of these topics in advanced calculus, these chapters may be considered as mainly review material and covered rapidly in class with emphasis on the later sections. A teacher who does this and also omits chapter six may be able to include most of the remaining parts of the book in a one-term course. With such an abbreviated course, the student of engineering or physics would study a consistent chain of mathematical theorems from the chapters covered in full, and from such a sample gain some idea of a mathematical science. The student majoring in mathematics might be stimulated to undertake a study of the rest of the material by himself.

A great many problems of varying difficulty are included in the forty exercises. For the more difficult problems, some indication of the solution is given as a *hint*. Where a problem requires algebraic or numerical answers, these are either included in the statement of the problem, or given immediately following the problem.

A short list of works, some recent publications and some references standard for many years, will be found in the bibliography. This list is by no means exhaustive; the titles are to be regarded as possible suggestions for further reading on this very extensive branch of mathematics.

Phillip Franklin

Cambridge, Mass.

Preface To The 2018 Edition

This is a basic textbook on complex analysis.

That's right, *another* textbook on complex analysis.

In academic mathematics, most people feel we need another complex analysis textbook like we need to have another lobbyist in Washington.

Or another deep dish pizza place in Chicago.

Or another chili place in Cincinnati.

Or another new IPhone put out by Apple.

Or another racist, sexist conservative running for office in the Deep South.

Or another *Fast And Furious* movie, when you're still trying to figure out why people spent money to go see the ones before it.

Or another Hilary Clinton presidential run.

(And that's just because of how it ended. At this point, it doesn't matter what her policies are. If if you run this corruptly arrogant, plutocratic and self-serving a campaign and your loss produces an outcome with this many horrific consequences for the entire planet, you and your entire family should be disqualified for the next 500 years from holding public office. Assuming of course, the world survives that long…………..)

Or another mass shooting with a military grade weapons that people yell thoughts and prayers for victims they don't give a shit about-which they prove by not doing anything significant to stop the next one.

(Ok, maybe that last one went a little too far, scratch that one……….)

My point is that the number of complex analysis textbooks that are currently available-both current publications and old classics-is so large, one is tempted to say it's uncountable. So why would you write a new original one, much less reach back almost 60 years to find one for republication?!?

Well, people continue to write new ones and republish out of print ones regardless. I can't speak for the motivations of those authors-all I can do is speak for myself and why I reached back into hallowed antiquity to make this one available again.

Well, one reason is while there are an immense number-a *ludicrous* number, frankly-of complex variables textbooks on the market, there are still relatively few standard ones that are available inexpensively. All you need to prove this is to look at what the standard undergraduate and graduate textbooks in use at the top schools on the subject are going for brand new at this writing on Amazon. The standard undergraduate textbooks-Brown/Churchill's *Complex Variables With*

Applications (9th edition, 2013,McGraw-Hill) and Saff/Snider's *Fundamentals of Complex Analysis: with Applications to Engineering and Science* (3rd edition, Pearson,2017) are $ 105 and 99 USD respectively. The standard graduate textbooks-Lars Alfhors' *Complex Analysis* (3rd edition, McGraw-Hill, 1979) or Walter Rudin's *Real And Complex Analysis* (3rd edition, McGraw Hill, 1987) are even more expensive, $159 and 132 respectively. Ok, granted today you can either get international softcover editions of these books that are much cheaper or choose other books. But despite being much cheaper than their American hardcover counterparts, international editions have other drawbacks, such as notorious fragility and off-times, being organized significantly differently from the originals. Also, let's face it-many departments would hesitate to use other books then the standard ones, particularly at top programs, due to an irrational, almost religious reverence for such works. Whether or not you agree with this reverence or not, it's ludicrously unfair that this decision has to be a problem at all because of the insane prices of these texts. Therefore I'm republishing this old classic because it's comparable to any of these in coverage and superior to some in exposition-and it'll be far less expensive.

Another reason is the wonderful pedigree of the author that's been all but forgotten. If you're a student or professor of mathematics and you' Phillip Franklin, you can be forgiven as his era now lies far behind the current generation. He was one of the most famous American-born researchers and teachers of mathematics in the first half of the 20th century. When I was an undergraduate, the older professors who were students at The Massachusetts Institute of Technology in the 1960's-like my mentor Nick Metas and real and functional analysis guru Gerald Itzkowitz-spoke of him with hushed awe when they met the then-long retired Franklin and told us how devastated the department was when he passed away suddenly. Franklin was one of the first American mathematicians with a truly modern training; getting his doctorate at Princeton University under Oswald Velben in 1922 and beginning his career by spreading knowledge of the frontiers of mathematics to the other top programs. Marshall Stone famously commented that the entire Harvard University department-students and faculty alike- learned the elements of topology from the seminar Franklin taught when he got there. He then moved on to MIT where he spent the bulk of his career, both teaching and doing research. He's best remembered for his research in analysis, but the truth is Franklin published in a number of fields, including number theory, graph theory and topology. A nice summary of how Franklin was perceived and remembered upon his death was given by Dirk Struik: *Phil was an even-tempered, mild, humorous man, "of almost Mr Chips proportions", as Dean Harrison said in* 1965 *at his funeral service……. He was well-versed in many fields of geometry, algebra and analysis."* We won't give a full biography of Franklin here, but much more about his fascinating life can be found at (1).

Like for most famous mathematicians, however, Rota's Dictum applies.

What? You never heard of Rota's Dictum?

There's a pretty good reason for that-I just made it up.

Well, I made up the name for it, anyway.

Rota's Dictum refers to the following quote by MIT's famous combinatorics guru Gian-Carlo Rota: *You are most likely to be remembered for your expository works.* (2) Its' easy to see why this is true: no matter how prolific a researcher is, the overwhelming majority of his or her papers will be incomprehensible to all but specialists. Textbooks, on the other hand, can reach a much larger audience.

Franklin was no different-and he authored some of the most well-received and influential textbooks of his generation. Among his most famous texts were *Differential equations for electrical engineers* (1933), the exhaustive basic analysis text *Treatise on advanced calculus* (1940), which quickly became one of the classic standard texts for advanced calculus in the pre-Rudin era, its companion sequel for physics and engineering students, *Methods of advanced calculus* (1944), the standard introductory calculus text *Differential and integral calculus* (1953) and the honors calculus text *Compact calculus* (1963).

Happily, several of them are now being republished for a new generation-the book you hold in your hand included. *Functions of a complex variable* (1958), which forms the heart of this book, was one of Franklin's last-and best-textbooks.

Yet another reason is the book's style: It is modern without being overly terse. Franklin's book is brief without lacking detail-it covers just about everything one wants covered in a first course on the subject and it does it in a brisk span of 255 pages. It emphasizes the wonderful interplay of analysis and geometry, that we'll describe a bit later, that forms the heart of classical complex analysis. The book also does so with minimal background: The only real prerequisites for most of the book are ordinary calculus up to and including multivariable calculus and a good course of "epsilon-delta" single variable rigorous analysis. Chapters 2-10 of our recently republished book by W.Benjamin Fite (3) will supply more than sufficient background for Franklin's book. For the sections on complex integration, experience with line integrals in vector analysis of the plane, such as can be found in chapter 3 of Kenneth Miller's book(4) would be quite helpful. It can be used for either a full year or intensive one semester undergraduate course in complex variables or for the first part of a first year graduate course that then is supplemented with more modern and sophisticated material. Because of the intertwining of geometry and analysis in complex variables, a good knowledge of classical plane geometry would be extremely helpful, but not really necessary.

(Indeed, I truly learned a great deal of classical geometry-which I never really learned in high school-in my first complex analysis course!)

Which brings me to the last and most important reason for republishing this textbook-simply that it's a hell of a good introductory textbook on complex analysis of one variable for students and the fact that it's out of print is a travesty. The very mission statement of Blue Collar Scholar Books is to rescue from publishing purgatory precisely such wonderful out of print jewels and make them available cheaply and widely to all students and teachers of the subject.

••

Complex analysis/complex variables, as the name for this major branch of analysis, has always struck me as a bit of an oxymoron. Seriously, no lie. The reason for this is because working in the complex plane and its generalizations, if anything *simplifies* many aspects of real analysis.

Walter Rudin, in the preface of the first edition of his famous-or infamous, depending on how much you suffered through it-graduate level analysis text *Real And Complex Analysis*, made a strikingly accurate observation comparing real and complex analysis:

Traditionally (with some oversimplification) the first of these deals with Lebesgue integration, with various types of convergence, and with the pathologies exhibited by very discontinuous functions; whereas the second one concerns itself only with those functions that are as smooth as can be, namely, the holomorphic ones.

Overall, more than "some" oversimplification is needed to make this description-but when one is discussing only the basics of either with students, this is remarkably accurate. In real analysis, even if one limits oneself to the one-dimensional normed vector space of the real numbers **R**, we have an insane menagerie of classifications of relative "smoothness" of functions: uniformly continuous functions, semi-continuous functions, one sided limits, one sided continuity, functions of bounded variation, uniformly differentiable functions, functions differentiable almost everywhere and of course, Weierstrass' explicit example of a function that is continuous everywhere but differentiable nowhere. The bizarreness of the real number system isn't limited to the functions, either. There's a legion of strange subsets of **R**: the long line, Cantor spaces and other fractal subsets just to name a few. Of course, these 2 aspects of real analysis are equivalent in many ways: Consider the Cantor set's functional counterpart, The Devil's Staircase. Entire texts have been written on the wild oscillations, infinite discontinuities and anomalous real valued functions defined on subspaces of $\mathbf{R}^n$.

To keep this simple, we'll limit the comparison to the functions defined in **R**, $\mathbf{R}^2$ and functions defined on **C**.

Nothing like this exists when functions are defined in the complex plane. Intuitively, this actually makes sense. Since negative numbers have square roots when one admits complex numbers, there are many points which are defined in both the domains and images of complex valued functions which do not exist in those of real valued functions. More precisely, the smoothness of functional behavior in the complex plane derives from complex differentiability, which is a much stronger condition then real differentiability. Most of the pathologies of functions in real analysis are derived from the fact that most functions have a finite order of differentiability-that is to say, the continuity of nth order derivatives or partial derivatives only exist up to a certain integer k. For example, C^2 functions in **R** have up to continuous second order derivatives.

But with complex analysis, differentiability is a far stronger condition. In fact, *a complex function is holomorphic (i.e. differentiable) if and only if it's infinitely differentiable on its domain.* A great deal of basic complex analysis is spent understanding why this is true. Such functions are called analytic functions. This absolute condition for differentiability in **C**, as you would imagine, has very profound consequences.

I should be careful in making clear that the equivalence of complex differentiability and analyticity only holds for complex valued functions. There are indeed real valued analytic functions-*but in the real case, analyticity does not necessarily imply infinite differentiability!* As one progresses in analysis, one finds this distinction is of paramount importance.

My point is that there is nothing like the vast spectrum of "smoothness types" of functions that one sees in real-valued functions or vector valued functions in real Euclidean spaces in functions defined in the complex plane.

That's not to say complex valued maps exhibit no pathologies-they would too simple to be interesting in that case. But the pathologies of complex maps are far more homogeneous and well behaved then those in real spaces. The main pathologies of complex maps are singularities or poles i.e. points where the function is undefined. These singularities can be assigned a complex number, called a residue,that is significant in integration in **C**. Complex valued functions which have isolated singularities but are analytic on all other points of their domain are called meromorphic functions. It's important to understand meromorphic functions are holomorphic everywhere they are defined. Indeed, whatever discontinuities do exist in images of complex valued functions of one variable, there are usually conditions under which they can be completely removed, rendering the function on the new domain or curve holomorphic i.e. complex differentiable! It is the study of these "nonsmooth" points that establishes most of the more sophisticated analytic tools of complex analysis.

Also, since each z in **C**, unlike real valued functions, is defined by an ordered pair (x,y) in the Cartesian plane, the domain and image of any function f: $\mathbf{C} \to \mathbf{C}$ defines "regions" of the complex plane bounded by "curves". Since every complex point is a point in the plane, the complex plane can be described very easily and clearly using polar coordinates-which leads to every complex number or complex valued mapping having a polar formulation.

The result is what is the single most important distinction between analysis on the real line and the complex plane: Since angles in the polar form of complex numbers are oriented i.e. they have positive or negative direction, this means that functions in the complex plane allow us to easily map regions in a one to one correspondence that is conformal or angle preserving. This has many deep consequences, but the most immediate and one of the most powerful consequences for the beginner is that many of the classical mappings and theorems of classical Euclidean and non-Euclidean geometries have very simple formulations and proofs when expressed in the complex plane by analytic mappings!

I've gone on probably too long here in my attempt to convince the prospective reader that complex analysis is a wonderful subject well worth pursuing. The quotations here are not accidental as these imprecise terms, while their meanings are intuitively clear here, must be precisely formulated for this to make mathematical sense. The result is that said reader is probably scratching his or her head right now in confusion.

Relax-all will be made clear in the course of this book. You will be in the extraordinarily capable hands of a master in Franklin. When the course is over or you're done working through this marvelous book on your own, you will have experienced all the basics of the remarkable branch of mathematics called complex analysis. And you'll understand why this beautiful subject has fascinated, thrilled and seduced both students and practitioners of both pure and applied mathematics for over 2 centuries.

We now give a brief description of the contents of the book. Chapter 1 introduces the complex numbers and the complex plane, describing the algebra and topology of the plane and how these relate in defining their most basic properties. Chapter 2 introduces functions of one complex plane and their calculus: limits, differentiation and some applications to geometry and analysis such as the solution of the Laplace equation. Chapter 3 extends power series approximations and their convergence properties to complex valued functions. Chapter 4 discusses the so-called elementary functions that form the building blocks of complex analysis; the exponential, the trigonometric functions, hyperbolic functions and power functions. It also contains an introduction to Riemann surfaces, which is such a critical construct not only in applications of complex analysis but it's generalization beyond one variable. Chapters 5 and 6 discuss the conformal mappings, which completely characterize the deep interplay of plane geometry and analysis with the complex numbers. In particular, the fractional bilinear transformation is examined in detail, which is one of the most important conformal mappings because it completely encodes all the classical Euclidean transformations of the plane. Chapter 7 discusses integration in the complex plane and the Cauchy-Goursat theorem which is the complex analogue of the fundamental theorem of calculus for path integrals in **C**. Chapter 8 discusses the Taylor series expansion in the complex plane and its many applications to analytic functions including Louisville's and Morera's theorems. Chapter 9 discusses the generalization of Taylor's formula with remainder to **C**, the Laurent series approximation formula. This brings us to the singularities of mappings and the coefficients of the local Laurent expansions, which are the residues. The final chapter describes many of the applications of residues in complex analysis, particularly in calculating the contour integrals of various meremorphic functions.

An interesting and important observation about the structure of the book related to the description of complex analysis given at the outset. While Franklin does balance the geometric and analytic aspects of the subject while not giving any priority over either-something you don't usually see in most texts on the subject, as you'll see in the recommended reading section- he does make a clear effort to linearly order the topics of the book on what he perceives as an axis of relative difficulty: The book's contents become steadily more difficult as one proceeds-and it's clear this implies the analytic aspects are back-loaded into the second half of the book. I think this is a very wise organizational approach, born of many years teaching the subject to very strong students in both pure and applied mathematics.

A word on the exercises in this book. One of the reasons Franklin can present such a thorough course in a relatively short book is that many of the easier results are shunted to the exercises, leaving the more difficult concepts for the text itself. It's here that I think it's appropriate to raise the famous adage about mathematics by the legendary problem-solver George Polya: *Mathematics is not a spectator sport.* While I think it's educational malpractice to put the building and development of significant tracts of theory that are required for understanding a new area of mathematics entirely in the exercises of a mathematics textbook, I think it's *student* malpractice to skip many of the exercises. Working exercises and problems is one of the most important steps-indeed, many believe ***the*** most important step-of the mathematical learning process. In this case, since many side issues and lemmas are left to the exercises, I believe it is more important than usual. The command you acquire of the subject will be much poorer if you just read the text and don't at least attempt all the exercises. So my advice to you is quite simple and direct: *Do as many of the exercises as time allows*. Besides-you paid for all of them, why not try them all?

I've babbled on more than long enough here trying to convince you that you made the right decision to learn complex analysis. But you know, it's like when a cop in the classic movie thriller *Heat* tries to convince Ashley Judd's character to flip on her criminal husband and his crew of professional thieves.

"What else are you selling?" she asks.

The cop replies, "All kinds of s***. *But you know I don't have to sell this s*** because this s*** sells itself.*"

I don't have to sell you on complex analysis because this s*** sells itself. And I hope after this course, you'll agree.

One last thing: I was first indoctrinated into the ways of complex functions by Ravi Kulkarni, the now-retired geometer at Queens College of the City University of New York in his deep and intensive undergraduate complex analysis course. His beautiful and detailed notes, which emphasized the geometric approach to the complex plane and the remarkable connections between the similarities and isometries of classical geometry and conformal mappings, made a sledgehammer impression on me at a time when I was reconsidering my academic shift from chemistry to mathematics. The passion, humor and warm patience and accessibility that he imbued every class he taught, particularly this one and every discussion he had with students, visitors and other faculty, cast the die for me in favor of The Queens of Sciences, Mathematics. There's a big hole at my alma mater where he used to be.

After 40 years teaching and doing research in geometry and complex analysis at John Hopkins, Indiana University and of course, my alma mater where I knew him, The City University of New York, Ravi returned to his birthplace of India where he's finally gotten the attention he deserves in the twilight of his career. He's been Distinguished Professor and Director of Harish-Chandra Research Institute, one of three research institutes for Mathematics and Theoretical Physics in India, followed by a 7-year stint at the Indian Institute of Technology (Bombay) as Mathematics Chair. He recently was honored for his long career at CUNY-and I'm so sorry to say I missed it.

Those who speak of Ravi now are surprised he's "recently" taken an interest in the philosophy of mathematics. Which of course, makes those of us who have been his students smirk. There's *nothing* recent about it-Ravi's *always* asked the deep questions about our science and encouraged his students to do so as well. Those who have read his photocopied notes on the 3 Aspects of Mathematics- Space, Number and Symmetry- are only surprised it took this long for other mathematicians to notice his interest in it. He was utterly fascinated by Felix Klein's *Erlangen* program and its context haunted all his lectures as he saw his own "3 Aspects" ideas as very much a sequel to Klein's program. My point is that Ravi has always been a metaphysician of the mathematical and always tried to get his students to think and question all things mathematical as well.

One of my most vivid memories of Ravi's metaphysical side occurred in his complex variables class. Before defining a function, he turned and asked the class what a function is. "Have any of you ever asked yourself what one is? You've all been using them for a good part of your life, but have any of you asked what it is? Can any of you give me a definition?" Someone innocently raised their hand and volunteered. "A function is a nonempty set of ordered pairs such that no different ordered pairs have the same first member." Before the student was halfway through his definition, Ravi's face soured and he yelled, "SHHHHHHHHHHHHHHHHHHHHHHHHHH!" Which of course elicited uproarious laughter from the student's classmates. Ravi's problem with this otherwise accurate set-theoretic definition is that it's completely devoid of any characteristics that we normally attribute to functions. "This definition would tell someone just learning mathematics *absolutely nothing*. It doesn't give them any ideas what functions do or why they play such a critical role in mathematics." I've done much hard thinking on the idea of mappings since that class-I'm not sure I agree with him. But I certainly see the point Ravi was trying to make.

Another time was when Ravi tried to get us to define an angle. "You're all supposed to know what angles are and you've used them most of your life. But do you really know what they are? Have any of you even asked yourselves?" Crickets. He proceeded to give us the following definition: "An angle is a function that assigns a real number *to a set of corners*." To which of course all of us looked at him like he had 3 eyes. I asked, "What do you mean by a set of corners?" Ravi smiled. "I'll let you think about that one. I'm more interested in assigning the numbers to them using the mapping."

That was Ravi.

One of my great academic regrets as a student was not being part of his legendary graduate course in differential geometry at the Graduate Center. With my own erratic health and my father's slow death, it just was never a good time. I did manage to get a photocopy of his lectures from a friend who did take his class-and it remains one of my most treasured possessions. And only deepens the regrets I wasn't actually part of it.

I hereby dedicate this text to my old Harvard-educated mentor, Ravi Kulkarni-and I hope those who read this tome will be half as inspired by it as Ravi inspired those of us lucky enough to learn under his tutelage.

This one's for you, Ravi. Consider it thanks from a student who never really thanked you enough.

Karo Maestro

New York City

March 2018

CONTENTS

1. Complex Numbers   1

1. Complex quantities, 1. – 2. Ordered pairs, 2. – 3. Operations on ordered pairs, 3. – 4. Notation, 4. – Exercise 1, 5. – 5. Complex conjugate, 5. – Exercise 2, 7. – 6. The complex plane, 8. – 7. The polar form, 9. – 8. Products and quotients in polar form, 11. – Exercise 3, 12. – 9. Powers and roots, 13. – Exercise 4, 15. – 10. Point sets, 16. – 11. The Heine-Borel theorem, 18. – Exercise 5, 18.

2. Functions of a Complex Variable   20

12. Functions, 20. – 13. Continuity, 21. – 14. Derivatives, 23. – 15. Differentiation rules, 24. – Exercise 6, 26. – 16. Cauchy-Riemann equations, 26. – Exercise 7, 29. – 17. Laplace's equation, 30. – Exercise 8, 32.

3. Power Series   35

18. Infinite series, 35. – 19. Power series, 37. – Exercise 9, 39. – 20. Uniform convergence, 40. – 21. Differentiation of power series, 43. – Exercise 10, 46.

4. The Elementary Functions   50

22. The exponential function, 50. – Exercise 11, 54. – 23. The trigonometric functions, 55. – Exercise 12, 59. – 24. The hyperbolic func-

CONTENTS

4. The Elementary Functions (Cont.)

tions, 61. – Exercise 13, 62. – 25. The logarithmic function, 64. – Exercise 14, 66. – 26. The general power function, 68. – 27. Inverse functions, 69. – 28. Elementary functions, 70. – 29. Riemann surfaces, 71. – Exercise 15, 73.

5. Conformal Transformations 77

30. Functions as transformations, 78. – 31. Continuously differentiable transformations, 79. – Exercise 16, 81. – 32. Conformal transformations, 83. – Exercise 17, 89. – 33. Boundary value problems, 91. – Exercise 18, 93.

6. Bilinear Transformations 98

34. Linear transformations, 99. – Exercise 19, 100. – 35. Bilinear transformations, 100. – 36. The transformation $w = 1/z$, 101. – 37. The point at infinity, 103. – Exercise 20, 104. – 38. Inversion in any "circle," 108. – 39. Equations of coaxial circles, 112. – Exercise 21, 113. – 40. Particular bilinear transformations, 116. – Exercise 22, 118.

7. Integral Theorems 121

41. Real definite integrals, 122. – 42. Curved paths of integration, 123. – 43. Integrals along curves, 124. – 44. Linear properties of integrals, 127. – Exercise 23, 128. – 45. Integral of a derivative, 129. – 46. An inequality for integrals, 130. – Exercise 24, 131. – 47. Cauchy's integral theorem, 132. – 48. The Cauchy-Goursat integral theorem for a triangle, 133. – 49. The Cauchy-Goursat integral theorem, 136. – Exercise 25, 138. – 50. Multiply connected regions, 139. – 51. The Cauchy integral formula, 140. – Exercise 26, 143.

8. Taylor's Expansion 148

52. Integration of uniformly convergent series, 149. – 53. Taylor's series, 149. – Exercise 27, 153. – 54. Morera's theorem, 159. – 55. Analytic functions, 160. – 56. Inequalities, 161. – 57. The maximum

CONTENTS

8. Taylor's Expansion (Cont.)

principle, 164. – 58. Liouville's theorem, 165. – Exercise 28, 167. – 59. Analytic continuation, 170. – 60. Definition by continuation, 172. – Exercise 29, 175.

9. Laurent's Expansion   179

61. Laurent's series, 180. – Exercise 30, 184. – 62. Singular points, 186. – 63. Rational functions, 192. – Exercise 31, 193. – 64. The residue theorem, 196. – 65. The argument principle, 198. – 66. Rouché's theorem, 200. – Exercise 32, 201. – 67. Residues of rational functions, 203. – 68. Conformal mapping of domains, 205. – 69. The Schwarz-Christoffel transformation, 206. – Exercise 33, 207.

0. Applications of Residues   209

70. Isolated singular points, 209. – 71. Residue at infinity, 210. – 72. Evaluation of residues, 211. – Exercise 34, 213. – 73. Real integrals found by using the unit circle, 213. – 74. Infinite integrals of rational functions, 215. – Exercise 35, 216. – 75. Infinite integrals with $\sin mx$ or $\cos mx$ as a factor, 218. – Exercise 36, 220. – 76. Principal value of an integral, 221. – 77. Indented contours, 222. – Exercise 37, 224. – 78. Integrals with a many-valued factor, 226. – 79. Special types of contour, 228. – Exercise 38, 229. – 80. Expansions in rational fractions, 232. – Exercise 39, 235. – 81. Summation of series, 236. – Exercise 40, 239.

Bibliography   241

Index   243

1

COMPLEX NUMBERS

The theory of functions of a complex variable plays a central role in both pure and applied mathematical analysis. A fundamental part of this theory deals with the application of the methods of the differential and integral calculus to complex numbers. As a preliminary to this, we devote this chapter to the study of complex numbers.

We define such numbers and the four fundamental operations of arithmetic for them. We describe their geometric interpretation, and conclude with a discussion of point sets.

1. Complex quantities

The need for complex numbers arises in algebra from the impossibility of finding, among the real numbers, square roots of negative quantities. Thus the equation $x^2 = -4$ has no real roots.

But we may invent an *imaginary unit* i for which

$$i^2 = -1. \tag{1}$$

We then define a *complex number* as a combination $a + bi$ formed from two real numbers a and b, and the imaginary unit i. We often omit terms with zero coefficients. And we do not write explicitly coefficients of i which are unity. Thus we write

$$a + 0i = a, \quad 0 + bi = bi, \quad 0 + 0i = 0, \qquad (2)$$

$$a + 1i = a + i, \quad a + (-1)i = a - i. \qquad (3)$$

These conventions make the system of complex numbers $a + bi$ include the real numbers a, as well as the pure imaginaries bi, and in particular, the imaginary unit i itself.

The laws of operation for complex numbers are taken as the ordinary laws of algebra for real quantities, with Eq. (1) added. These rules provide $x = 2i$ or $x = -2i$ as the two roots of $x^2 = -4$. And the formula $x = (-B \pm \sqrt{B^2 - 4AC})/2A$ then provides two (possibly coincident) complex roots for every quadratic equation $Ax^2 + Bx + C = 0$ with $A \neq 0$.

We have sketched this traditional approach to the complex number system to relate our subject to the reader's earlier study. But we shall describe an alternative method of defining complex numbers in terms of real numbers. And we shall regard this more abstract procedure as the basis of all later definitions and arguments.

2. Ordered pairs

Definition. A complex number is an ordered pair of real numbers. Thus $(\frac{1}{2}, -2)$, $(1, \sqrt{3})$ are complex numbers. We often write $z = (x,y)$. The real number x is called the *real component* of the complex number z. And the real number y is called the *imaginary component* of the complex number z. For these components we shall use the notation

$$\text{Re } z = \text{Re}(z) = x, \quad \text{Im } z = \text{Im}(z) = y. \qquad (4)$$

The absolute value of the complex number z is the real number $+\sqrt{x^2 + y^2}$. We denote it by vertical bars, so that

$$|z| = +\sqrt{x^2 + y^2}. \qquad (5)$$

Whenever x or y differs from zero, $|z|$ is positive. Hence

$$|z| = 0 \quad \text{implies that:} \quad x = 0 \text{ and } y = 0. \qquad (6)$$

We do not establish any linear order for complex numbers. Consequently we do not apply the relations "greater than" or "less than" to complex numbers without a further qualification. But inequalities such as $|z_2| > |z_1|$ may apply to the real absolute values of two complex numbers.

Two complex numbers are equal if and only if their real and imaginary components are separately equal. The equation

$$(x_1, y_1) = (x_2, y_2) \quad \text{implies that:} \quad x_1 = x_2 \text{ and } y_1 = y_2. \tag{7}$$

3. Operations on ordered pairs

Let $z_1 = (x_1, y_1)$ and $z_2 = (x_2, y_2)$ be any two complex numbers. Then the equation

$$z_1 + z_2 = (x_1 + x_2, y_1 + y_2) \tag{8}$$

defines the operation of *addition*, or of finding the sum of two complex numbers. It follows from Eq. (8) that the *commutative* law of addition,

$$z_1 + z_2 = z_2 + z_1, \tag{9}$$

is satisfied. And the *associative* law of addition

$$z_1 + (z_2 + z_3) = (z_1 + z_2) + z_3 = z_1 + z_2 + z_3, \tag{10}$$

also holds.

The operation of *multiplication*, or of finding the product of two complex numbers is *defined* by the equation

$$z_1 z_2 = (x_1 x_2 - y_1 y_2, x_1 y_2 + x_2 y_1). \tag{11}$$

It follows from Eq. (11) that the commutative law of multiplication,

$$z_1 z_2 = z_2 z_1, \tag{12}$$

is satisfied. And the associative law of multiplication

$$z_1(z_2 z_3) = (z_1 z_2) z_3 = z_1 z_2 z_3, \tag{13}$$

also holds.

The *distributive* law of multiplication with respect to addition,

$$z_1(z_2 + z_3) = z_1 z_2 + z_1 z_3, \tag{14}$$

is another consequence of Eqs. (8) and (11).

The operation of *subtraction* is *defined* by the equation

$$z_1 - z_2 = (x_1 - x_2, y_1 - y_2). \tag{15}$$

This makes subtraction the inverse of addition in the sense that if $z_1 - z_2 = z_3$, then $z_1 = z_3 + z_2$.

The operation of *division* is *defined* by the equation

$$\frac{z_1}{z_2} = \left(\frac{x_1 x_2 + y_1 y_2}{x_2^2 + y_2^2}, \frac{x_2 y_1 - x_1 y_2}{x_2^2 + y_2^2} \right), \quad (16)$$

if $|z_2| \neq 0$. Division by $(0,0)$ is not defined. This makes division the inverse of multiplication in the sense that if $z_1/z_2 = z_3$, then $z_1 = z_3 z_2$.

The four fundamental operations of algebra are addition, subtraction, multiplication, and division. Thus Eqs. (8), (15), (11), and (16) define an algebra of complex numbers. And the basic laws of this algebra are the same as those of the familiar algebra of real numbers. Hence when a single letter is used to denote a complex number, the two algebras will be formally the same. But since the real algebra deals with single numbers, while the complex algebra deals with ordered pairs of real numbers, the interpretations will be different.

4. Notation

Let us use x as an abbreviation for the complex number $(x,0)$ with imaginary component zero. Then from Eq. (8) it follows that

$$x_1 + x_2 = (x_1, 0) + (x_2, 0) = (x_1 + x_2, 0). \quad (17)$$

From Eq. (11) it follows that

$$x_1 x_2 = (x_1, 0)(x_2, 0) = (x_1 x_2, 0). \quad (18)$$

This shows that treating complex numbers with imaginary component zero as though they were real numbers is consistent with our rules of operation for ordered pairs.

Again, let us use i to denote the particular ordered pair $(0,1)$. Then it follows from Eq. (11) that

$$i^2 = (0,1)(0,1) = (-1,0) = -1. \quad (19)$$

Thus $i = (0,1)$ has the fundamental property of the imaginary unit stated in Eq. (1).

It also follows from Eq. (11) that

$$iy = (0,1)(y,0) = (0,y). \quad (20)$$

Hence from Eq. (8), we may deduce that

$$x + iy = (x,0) + (0,y) = (x,y). \quad (21)$$

Thus we may use $x + iy$ to denote the ordered pair (x,y). It is usually more convenient to carry out algebraic manipulations on this form instead of going back to the definitions of Section 3. We rewrite Eq. (7) in the form

$$x_1 + iy_1 = x_2 + iy_2 \quad \text{implies that} \quad x_1 = x_2 \text{ and } y_1 = y_2. \quad (22)$$

Exercise 1

Verify each of the following results:

1. $(-2 - 5i) + (3 + 4i) = 1 - i$.
2. $(6 - i)(6 + i) = 37$.
3. $(2 + 4i) - (3 - i) = -1 + 5i$.
4. $(2 - 2i)^2 = -8i$.
5. $z^2 - 6z + 13 = 0$ if $z = 3 \pm 2i$.
6. Re $(4 - 3i) = 4$.
7. Im $(4 - 3i) = -3$.
8. $|4 - 3i| = 5$.
9. Re $(5i) = 0$. 10. Im $(5i) = 5$. 11. $|5i| = 5$.

12. For any $z = x + iy$, show that Re $(-iz) = \text{Im } z = y$, and that Im $(iz) = \text{Re } z = x$.

13. Evaluate $(z_1 z_2)z_3 = [(x_1 + iy_1)(x_2 + iy_2)](x_3 + iy_3)$, and deduce from the symmetry of the result in the subscripts that $(z_1 z_2)z_3 = z_1(z_2 z_3) = z_2(z_1 z_3)$.

14. If $x_1 + iy_1 = (x + iy) + (x_2 + iy_2)$, deduce that $x_1 = x + x_2$ and $y_1 = y + y_2$. Solve these equations for x and y, and thus show that Eq. (15) is the only definition of subtraction which makes it the inverse of addition.

15. If $x_1 + iy_1 = (x + iy)(x_2 + iy_2)$, deduce that $x_1 = xx_2 - yy_2$ and $y_1 = xy_2 + yx_2$. Solve these equations for x and y, and thus show that Eq. (16) is the only definition of division which makes it the inverse of multiplication.

5. Complex conjugate

The *complex conjugate* of the complex number $z = x + iy$ is $x - iy$. We denote it by $\bar{z}$. It follows from this and the definitions of Section 3 that the conjugate of $z_1 + z_2$ is $\bar{z}_1 + \bar{z}_2$, of $z_1 - z_2$ is $\bar{z}_1 - \bar{z}_2$, of $z_1 z_2$ is $\bar{z}_1 \bar{z}_2$, and of z_1/z_2 is $\bar{z}_1/\bar{z}_2$. Consequently, *any equation between complex numbers which involves only the operations of addition, subtraction, multiplication, and division remains true if all the complex numbers which appear in the equation are replaced by their complex conjugates.*

The complex conjugate of $\bar{z}$ is z itself.

Since $z = x + iy$, $\bar{z} = x - iy$, we have from Eqs. (11) and (5),

$$z\bar{z} = x^2 + y^2 = |z|^2, \qquad (23)$$

and from Eqs. (9), (15), and (4) we may deduce that

$$z + \bar{z} = 2x = 2\,\mathrm{Re}\,z, \quad z - \bar{z} = 2iy = 2i\,\mathrm{Im}\,z. \qquad (24)$$

We illustrate the use of these relations by proving certain properties of absolute values. From Eq. (23) with $z = z_1 z_2$ we have

$$|z_1 z_2|^2 = (z_1 z_2)(\overline{z_1 z_2}) = z_1 z_2 \bar{z}_1 \bar{z}_2 = (z_1 \bar{z}_1)(z_2 \bar{z}_2)$$
$$= |z_1|^2 |z_2|^2. \qquad (25)$$

Taking the positive square roots of both sides leads to

$$|z_1 z_2| = |z_1||z_2|. \qquad (26)$$

That is: *The absolute value of the product of two complex numbers is the product of their absolute values.*

We may deduce from Eqs. (26) and (6) that

$$z_1 z_2 = 0 \quad \text{implies} \quad z_1 = 0 \text{ or } z_2 = 0. \qquad (27)$$

It follows from Eq. (5) that $x^2 \leq |z|^2$, $y^2 \leq |z|^2$, so that

$$|\mathrm{Re}\,z| \leq |z| \quad \text{and} \quad |\mathrm{Im}\,z| \leq |z|. \qquad (28)$$

We shall derive two other inequalities involving absolute values. From Eq. (23) with $z = z_1 + z_2$ we have

$$|z_1 + z_2|^2 = (z_1 + z_2)(\bar{z}_1 + \bar{z}_2)$$
$$= z_1 \bar{z}_1 + z_1 \bar{z}_2 + z_2 \bar{z}_1 + z_2 \bar{z}_2. \qquad (29)$$

But by Eqs. (24) and (28) with $z = z_1 \bar{z}_2$, $\bar{z} = \bar{z}_1 z_2 = z_2 \bar{z}_1$,

$$z_1 \bar{z}_2 + z_2 \bar{z}_1 = 2\,\mathrm{Re}\,(z_1 \bar{z}_2) \leq 2|z_1 \bar{z}_2| \quad \text{or} \quad 2|z_1||z_2|. \qquad (30)$$

It follows from Eqs. (29) and (30) that

$$|z_1 + z_2|^2 \leq |z_1|^2 + 2|z_1||z_2| + |z_2|^2 \quad \text{or} \quad (|z_1| + |z_2|)^2. \qquad (31)$$

Taking the positive square roots of both sides leads to

$$|z_1 + z_2| \leq |z_1| + |z_2|. \qquad (32)$$

That is: *The absolute value of the sum of two complex numbers cannot exceed the sum of their absolute values.*

By induction, it follows from Eq. (32) that for any n complex numbers z_q, a similar result holds, so that

$$|z_1 + z_2 + \ldots + z_n| \leq |z_1| + |z_2| + \ldots + |z_n|. \tag{33}$$

Again, from Eq. (23) with $z = z_1 - z_2$ we have

$$|z_1 - z_2|^2 = (z_1 - z_2)(\bar{z}_1 - \bar{z}_2) = z_1\bar{z}_1 - z_1\bar{z}_2 - z_2\bar{z}_1 + z_2\bar{z}_2. \tag{34}$$

It follows from this and Eq. (30) that

$$|z_1 - z_2|^2 \geq |z_1|^2 - 2|z_1||z_2| + |z_2|^2 \quad \text{or} \quad (|z_1| - |z_2|)^2. \tag{35}$$

Taking the positive square roots of both sides leads to

$$|z_1 - z_2| \geq ||z_1| - |z_2||. \tag{36}$$

We may also use conjugate complex numbers to simplify the calculation of quotients. Thus

$$\frac{z_1}{z_2} = \frac{z_1\bar{z}_2}{z_2\bar{z}_2} = \frac{(x_1 + iy_1)(x_2 - iy_2)}{|x_2 + iy_2|^2}$$

$$= \frac{(x_1x_2 - y_1y_2) + i(x_2y_1 - x_1y_2)}{x_2^2 + y_2^2}, \tag{37}$$

in agreement with Eq. (16).

Exercise 2

Verify each of the following calculations.

1. $\dfrac{2 + 6i}{1 - i} = \dfrac{(2 + 6i)(1 + i)}{|1 - i|^2} = -2 + 4i$.

2. $\dfrac{3 - 4i}{1 + 2i} = \dfrac{(3 - 4i)(1 - 2i)}{|1 + 2i|^2} = -1 - 2i$.

3. $\dfrac{3 + i}{-5 - 5i} = \dfrac{(3 + i)(1 - i)}{-5|1 + i|^2} = \dfrac{-2}{5} + \dfrac{i}{5}$.

Show by a direct calculation that

4. The conjugate of $(z_1 + z_2 - z_3)$ is $(\bar{z}_1 + \bar{z}_2 - \bar{z}_3)$.

5. The conjugate of $(z_1z_2 + z_1z_3)$ is $(\bar{z}_1\bar{z}_2 + \bar{z}_1\bar{z}_3)$.

6. $(x_1x_2 - y_1y_2)^2 + (x_1y_2 + x_2y_1)^2 = (x_1^2 + y_1^2)(x_2^2 + y_2^2)$, which checks Eq. (26).

7. Prove that $z = \bar{z}$ if and only if Im $z = 0$, so that $z = x = (x,0)$ is a *real number*.

8. Prove that $z = -\bar{z}$ if and only if Re $z = 0$, so that $z = iy = (0,y)$ is a *pure imaginary number*.

9. Prove that for any complex number $z + \bar{z}$ and $z\bar{z}$ are both real numbers.

10. Prove that if $z + z_1$ and $z\bar{z}_1$ are both real, either z and z_1 are both real, or $z_1 = \bar{z}$.

11. Verify that, if $z_2 \neq 0$,
$$\left|\frac{z_1}{z_2}\right| = \frac{|z_1\bar{z}_2|}{|z_2\bar{z}_2|} = \frac{|z_1||z_2|}{|z_2|^2} = \frac{|z_1|}{|z_2|}.$$

12. Prove the result of Problem 11, if $z_2 \neq 0$,
$$\left|\frac{z_1}{z_2}\right| = \frac{|z_1|}{|z_2|},$$
by applying Eq. (26) to the product of (z_1/z_2) and z_2.

13. From $2x^2 + 2y^2 = (|x| + |y|)^2 + (|x| - |y|)^2$, deduce that
$$|z| \geq (1/\sqrt{2})(|x| + |y|).$$

14. Prove that $|z_1 + z_2| \geq ||z_1| - |z_2||$.

15. Prove that $z + 1/z$ is real only if Im $z = 0$, or if $|z| = 1$.

16. Prove that
$$|z_1 + z_2|^2 + |z_1 - z_2|^2 = 2|z_1|^2 + 2|z_2|^2.$$

6. The complex plane

The complex number $z = x + iy$ is uniquely determined by the ordered pair of real numbers (x,y). The same is true of the point in the plane $P(x,y)$ with Cartesian coordinates x and y. Hence it is possible to establish a one-to-one correspondence between all the complex numbers and all the points of a given plane. We merely associate the complex number $x + iy$ with the point $P(x,y)$. For simplicity, we use z to denote the point as well as the complex number. The plane whose points represent the complex numbers is called the *complex plane* or the *z-plane*.

The *real numbers*, or points $x = (x,0)$ are represented by points on the *x*-axis, or axis of reals. And the *pure imaginary numbers*, or points $iy = (0,y)$ are represented by points on the y-axis, or axis of imaginaries. The complex number $0 = (0,0)$ is represented by the origin O. Instead of considering the point $P = (x,y)$ as the representation of $z = x + iy$, Fig. 1, we may equally well consider the directed segment or vector extending from the origin O to P as the representative of the complex number. In this case, any parallel segment of the same length and direction is taken as corresponding to the same complex number.

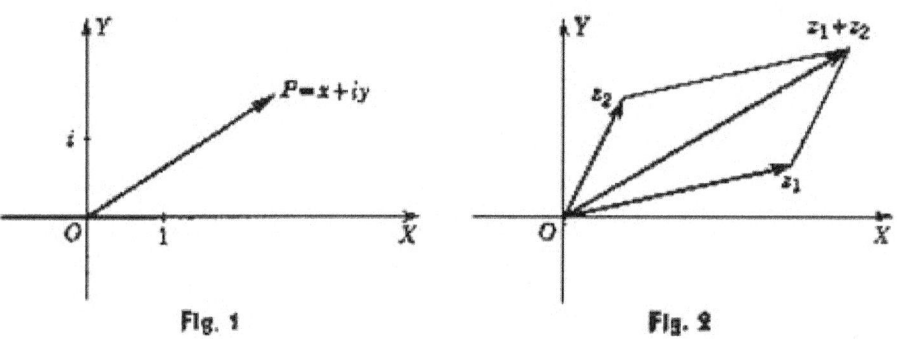

Fig. 1   Fig. 2

It follows from Eq. (8) that the vector which represents $z_1 + z_2$ can be found by adding the vectors which represent z_1 and z_2 graphically as indicated in Fig. 2. The vector for $z_1 - z_2$ can be found either

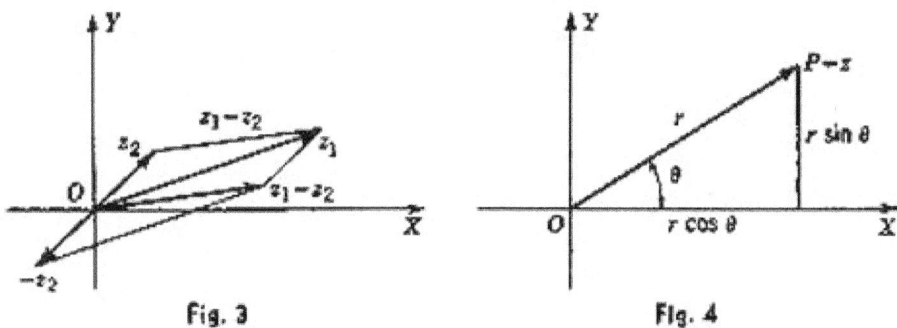

Fig. 3   Fig. 4

by adding the vector for z_1 to the vector for $-z_2$, or as the vector from the point z_2 to the point z_1. See Fig. 3.

7. The polar form

For the point z or $P(x,y)$ of Fig. 4, let us introduce polar coordinates r, θ. Thus r is the length of OP, and θ is an angle which rotates OX

into OP, considered positive when counterclockwise. Then the following relations hold:

$$x = r \cos \theta, \qquad y = r \sin \theta, \qquad (38)$$

$$r = \sqrt{x^2 + y^2}, \qquad \theta = \tan^{-1} \frac{y}{x}. \qquad (39)$$

When $z \neq 0$, we restrict r to its *positive* value. Hence in Eq. (39) we take the positive square root, and by Eq. (5), $r = |z|$, the absolute value of z. Also we must take a value of the inverse tangent in Eq. (39) for which Eq. (38) holds. Thus θ is in the same quadrant as $z = (x,y)$ and is determined to within an integral multiple of 2π radians if $z \neq 0$. When $z = 0$, $r = 0$ and θ may have any value. The angle θ is called the *argument* of z, and we write $\theta = \arg z$.

In view of Eq. (38), we may write

$$\begin{aligned} z = x + iy &= r \cos \theta + ir \sin \theta \\ &= r(\cos \theta + i \sin \theta). \end{aligned} \qquad (40)$$

This is the *polar form* of z. We shall use cis θ as an abbreviation for $\cos \theta + i \sin \theta$, so that

$$z = r \text{ cis } \theta = r(\cos \theta + i \sin \theta). \qquad (41)$$

We note that the complex conjugate of z is

$$\begin{aligned} \bar{z} = r \cos \theta - ir \sin \theta &= r[\cos (-\theta) + i \sin (-\theta)] \\ &= r \text{ cis } (-\theta). \end{aligned} \qquad (42)$$

Thus $\arg \bar{z}$ has $-\arg z$ as one value. As indicated in Fig. 5, the point $\bar{z}$ is the reflection of z in the x-axis, so that the segment from $\bar{z}$ to z has the x-axis as its perpendicular bisector.

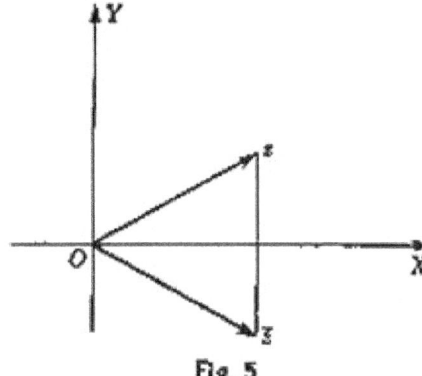

Fig. 5

8. Products and quotients in polar form

To find the product of the two complex numbers $z_1 = r_1 \operatorname{cis} \theta_1$ and $z_2 = r_2 \operatorname{cis} \theta_2$, we first calculate

$$\operatorname{cis} \theta_1 \operatorname{cis} \theta_2$$
$$= (\cos \theta_1 + i \sin \theta_1)(\cos \theta_2 + i \sin \theta_2)$$
$$= \cos \theta_1 \cos \theta_2 - \sin \theta_1 \sin \theta_2 + i(\sin \theta_1 \cos \theta_2 + \cos \theta_1 \sin \theta_2)$$
$$= \cos (\theta_1 + \theta_2) + i \sin (\theta_1 + \theta_2) = \operatorname{cis} (\theta_1 + \theta_2). \tag{43}$$

By multiplying in $r_1 r_2$, we may deduce from this that:

$$z_1 z_2 = (r_1 \operatorname{cis} \theta_1)(r_2 \operatorname{cis} \theta_2) = r_1 r_2 \operatorname{cis} (\theta_1 + \theta_2). \tag{44}$$

This shows that when two complex numbers in polar form are multiplied together, the *radius* of the product is the *product* of the radii of the factors, in accord with Eq. (26), and the *angle* of the product is the *sum* of the angles of the factors.

It follows from Eq. (44) by induction that

$$z_1 z_2 \ldots z_n = r_1 r_2 \ldots r_n \operatorname{cis} (\theta_1 + \theta_2 + \ldots + \theta_n). \tag{45}$$

To find the quotient of z_1 by z_2, we note that

$$\frac{z_1}{z_2} = \frac{r_1 \operatorname{cis} \theta_1}{r_2 \operatorname{cis} \theta_2} = \frac{r_1 \operatorname{cis} \theta_1 \operatorname{cis} (-\theta_2)}{r_2 \operatorname{cis} \theta_2 \operatorname{cis} (-\theta_2)}$$
$$= \frac{r_1 \operatorname{cis} (\theta_1 - \theta_2)}{r_2 \operatorname{cis} 0} = \frac{r_1}{r_2} \operatorname{cis} (\theta_1 - \theta_2),$$

since $\operatorname{cis} 0 = \cos 0 + i \sin 0 = 1$. Thus, when two complex numbers in polar form are divided, the *radius* of the quotient is the *quotient* of the radii, and the *angle* of the quotient is the difference of the angles. See Figs. 6 and 7 and Problems 16 and 17 of Exercise 3.

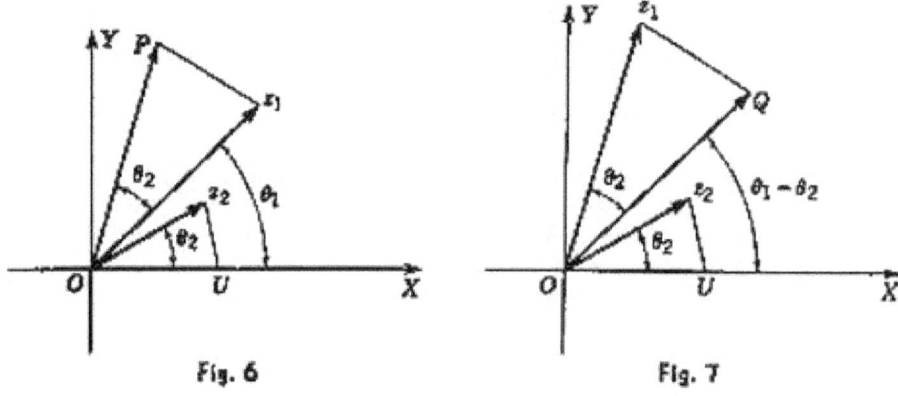

Fig. 6   Fig. 7

Exercise 3

Given $z_1 = 3 + 5i$, $z_2 = -5 + 3i$, $z_3 = 4i$, $z_4 = -2$, compute and check by a graph each of the following:

1. $z_1 + z_2$.
2. $z_3 + z_4$.
3. $z_1 - z_2$.
4. $z_2 - z_3$.

5. Show that the mid-point of the segment joining z_1 and z_2 is $z_M = \frac{1}{2}(z_1 + z_2)$.

6. Show that the point dividing the segment joining z_1 and z_2 in the real ratio s_1/s_2 is
$$z_D = z_1 + \frac{s_1}{s_1 + s_2}(z_2 - z_1) = \frac{s_2 z_1 + s_1 z_2}{s_1 + s_2}.$$

7. Show that $\frac{1}{3}(z_1 + z_2 + z_3)$ is the median point of the triangle whose vertices are z_1, z_2, and z_3.

Given $z_1 = 2\sqrt{3} - 2i = 4 \operatorname{cis}(-\pi/6)$, $z_2 = 1 + i\sqrt{3} = 2 \operatorname{cis} \pi/3$, $z_3 = 4 + 4i = 4\sqrt{2} \operatorname{cis} \pi/4$, $z_4 = -2 + 2i = 2\sqrt{2} \operatorname{cis} 3\pi/4$, use the polar form to evaluate each of the following:

8. $z_1 z_2$.
9. $z_3 z_4$.
10. z_1/z_2.
11. z_3/z_4.

12. Verify that Eq. (32) corresponds to the geometric property that the sum of any two sides of a triangle exceeds the third side. Deduce from this that the equality can hold only when $x_2 y_1 = x_1 y_2$.

13. Verify that Eq. (36) corresponds to the geometric property that any side of a triangle exceeds the difference of the other two sides. Deduce from this that the equality can hold only when $x_2 y_1 = x_1 y_2$.

14. Show that the equality sign holds in Eqs. (32) and (36) if $\arg z_1 = \arg z_2$. See Problems 12 and 13.

15. Consider a polygon of $n + 1$ sides whose first n sides are vectors representing $z_1, z_2, \ldots, z_n$. Then the closing side, in the same sense, represents $-(z_1 + z_2 + \ldots + z_n)$. Use this fact to deduce Eq. (33) from the geometric fact that a straight line is the shortest distance between two points.

In Figs. 6 and 7, O is the origin $(0,0)$ and U is the unit point $(1,0)$. Verify that the triangle with vertices OUz_2 is similar to the triangle with vertices.

16. $Oz_1 P$ if $P = z_1 z_2$.
17. OQz_1 if $Q = z_1/z_2$.

18. Let a, b, c be real quantities with $b^2 < 4ac$. Show that the two complex roots of $az^2 + bz + c = 0$ may be expressed in the form $r \operatorname{cis}(\pm\theta)$, where $r = \sqrt{c/a}$ and $\theta = \cos^{-1}(-b/2\sqrt{ac})$.

19. We say that the actual or degenerate similar triangles $z_1z_2z_3$ and $Z_1Z_2Z_3$ are directly similar if corresponding equal angles have the same sense. Derive the condition for this:

$$(z_1 - z_2)(Z_2 - Z_3) = (z_2 - z_3)(Z_1 - Z_2)$$

or

$$\begin{vmatrix} 1 & 1 & 1 \\ z_1 & z_2 & z_3 \\ Z_1 & Z_2 & Z_3 \end{vmatrix} = 0. \quad (46)$$

20. Show that the triangle $z_1z_2z_3$ is equilateral if and only if

$$z_1^2 + z_2^2 + z_3^2 = z_2z_3 + z_3z_1 + z_1z_2.$$

9. Powers and roots

By taking all the numbers z_1 to z_n equal in Eq. (45), we find that for any positive integer n,

$$z^n = (r \operatorname{cis} \theta)^n = r^n \operatorname{cis} n\theta$$
$$= r^n(\cos n\theta + i \sin n\theta). \quad (47)$$

That the same formula also holds for negative powers is seen from

$$z^{-n} = \frac{1}{z^n} = \frac{1 \operatorname{cis} 0}{r^n \operatorname{cis} n\theta} = \frac{1}{r^n} \operatorname{cis}(0 - n\theta) = r^{-n} \operatorname{cis}(-n\theta). \quad (48)$$

The result in Eqs. (47) and (48) is known as *De Moivre's theorem*.

In particular, from Eq. (48) with $n = 1$, we have

$$z^{-1} = \frac{1}{z} = \frac{1}{r} \operatorname{cis}(-\theta) = \frac{1}{r}(\cos\theta - i\sin\theta). \quad (49)$$

To find the nth root of z, we note that $z_1 = \sqrt[n]{z} = z^{1/n}$ if $z = z_1^n$. Thus if $z = r \operatorname{cis} \theta$, $z_1 = r_1 \operatorname{cis} \theta_1$, we have

$$r \operatorname{cis} \theta = (r_1 \operatorname{cis} \theta_1)^n = r_1^n \operatorname{cis} n\theta_1. \quad (50)$$

Consequently, $r_1^n = r$ and $n\theta_1 = \theta \pm 2k\pi$, where k is zero or a positive integer. Since r and r_1 are positive, it follows that $r_1 = \sqrt[n]{r}$, the arithmetic nth root of r. The values $\theta_1 = (\theta \pm 2k\pi)/n$ lead to the same value of z_1 for any two values of k that differ by a multiple of n. Thus there are exactly n distinct values of $z^{1/n}$, given by

$$z^{1/n} = \sqrt[n]{r}\operatorname{cis}\frac{\theta + 2k\pi}{n}, \qquad k = 0, 1, 2, \ldots, n-1. \tag{51}$$

As a particular case, let $z = 1$. Since $1 = 1 \operatorname{cis} 0$, $r = 1$, $\sqrt[n]{r} = 1$, and $\theta = 0$. Thus

$$1^{1/n} = \operatorname{cis}\frac{2k\pi}{n}, \qquad k = 0, 1, 2, \ldots, n-1. \tag{52}$$

If we denote $\operatorname{cis} 2\pi/n$ by ω, $\operatorname{cis} 2k\pi/n = \omega^k$, by Eq. (47). Thus the nth roots of unity may be written in the form

$$1, \quad \omega, \quad \omega^2, \quad \ldots, \quad \omega^{n-1} \quad \text{where} \quad \omega = \operatorname{cis}\frac{2\pi}{k}. \tag{53}$$

Since all the values of Eq. (51) may be obtained from any one by multiplication by $\operatorname{cis} 2k\pi/n = \omega^k$, it follows that if z_0 is any one nth root of z, all n roots are given by

$$z_0, \quad z_0\omega, \quad z_0\omega^2, \quad \ldots, \quad z_0\omega^{n-1}. \tag{54}$$

In the complex plane the roots of Eq. (54) or (53) all lie on a circle with center at the origin, and are the vertices of a regular polygon of

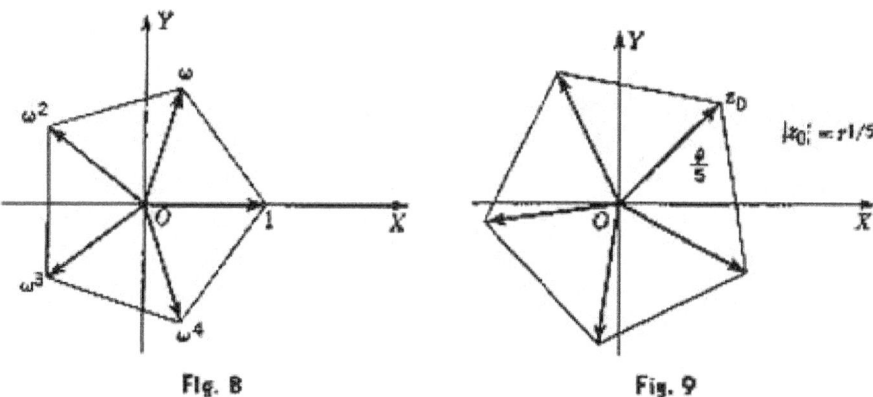

Fig. 8   Fig. 9

n sides. For Eq. (53), the radius of the circle is 1, and one vertex has argument zero, Fig. 8. For Eq. (54) or (51), the radius is $|z_0| = r^{1/n}$, and one vertex has argument θ/n, Fig. 9.

Exercise 4

Given $z_1 = -2 + 2i = 2^{3/2} \operatorname{cis} 3\pi/4$, verify that

1. $z_1^2 = -8i$. 2. $z_1^4 = -64$.
3. $z_1^{1/3} = 1 + i$, $\sqrt{2}(-\cos 15° + i \sin 15°)$, $\sqrt{2}(\sin 15° - i \cos 15°)$.

Verify that

4. $8^{1/3} = 2, -1 + \sqrt{3}i, -1 - \sqrt{3}i$.
5. $(-4)^{1/4} = 1 + i, -1 + i, -1 - i, 1 - i$.
6. From the fourth roots of -4, deduce the real factorization $z^4 + 4 = (z^2 + 2z + 2)(z^2 - 2z + 2)$.
7. From the sixth roots of 64, deduce the real factorization $z^6 - 64 = (z^2 - 4)(z^2 + 2z + 4)(z^2 - 2z + 4)$.
8. Show that the three cube roots of unity are 1, $\omega_1 = \dfrac{-1 + \sqrt{3}\,i}{2}$, $\omega_2 = \dfrac{-1 - \sqrt{3}\,i}{2}$. Also verify that $\omega_2 = \omega_1^2$, $\omega_1 = \omega_2^2$, $1 + \omega_1 + \omega_1^2 = 1 + \omega_2 + \omega_2^2 = 0$.

Use De Moivre's theorem to prove that:

9. $\cos 3\theta = \cos^3 \theta - 3 \cos \theta \sin^2 \theta$.
10. $\cos 4\theta = \cos^4 \theta - 6 \cos^2 \theta \sin^2 \theta + \sin^4 \theta$.
11. $\sin 3\theta = 3 \cos^2 \theta \sin \theta - \sin^3 \theta$.
12. $\sin 4\theta = 4 \cos^3 \theta \sin \theta - 4 \cos \theta \sin^3 \theta$.

13. Show that if $\omega = \operatorname{cis} 2\pi/n$, the vector ωz is the one obtained by rotating the vector z through an angle $2\pi/n$ in the positive direction without changing its length.

14. As the special case of Problem 13 with $n = 4$, show that iz is the vector z rotated through a right angle counterclockwise.

15. Show that if the median point of a triangle coincides with its circumcenter, then the triangle is equilateral. *Hint:* Take the median point at the origin, and let the vertices be z_1, z_2, z_3. Then by Problem 7, Ex. 3, $z_1 + z_2 + z_3 = 0$. Thus

$$z_1^2 + z_2^2 + z_3^2 = -2(z_2 z_3 + z_3 z_1 + z_1 z_2).$$

Also $\bar{z}_1 + \bar{z}_2 + \bar{z}_3 = 0$. But if $|z_1| = |z_2| = |z_3| = r$, $1/z_i = \bar{z}_i/r^2$ and $z_2 z_3 + z_3 z_1 + z_1 z_2 = z_1 z_2 z_3 (1/z_1 + 1/z_2 + 1/z_3) = z_1 z_2 z_3 (\bar{z}_1 + \bar{z}_2 + \bar{z}_3)/r^2 = 0$. Now use Problem 20 of Ex. 3.

16. Show that if a quadrilateral is inscribed in a circle whose center is the centroid of its four vertices, then the quadrilateral is a rectangle. *Hint:* As in the hint to Problem 15,

$$z_1 + z_2 + z_3 + z_4 = 0, \quad \sum z_2 z_3 z_4 = z_1 z_2 z_3 z_4 (\sum \bar{z}_1)/r^2 = 0.$$

Hence the polynomial $(z - z_1)(z - z_2)(z - z_3)(z - z_4) = z^4 + Az^2 + B$, so that with suitably numbered zeros, $z_3 = -z_1$ and $z_4 = -z_2$.

17. If m is an integer and n is a positive integer, show that

$$z^{m/n} = r^{m/n} \operatorname{cis} \frac{m}{n}(\theta + 2k\pi).$$

And if m and n have no common factors, the values $k = 0, 1, 2, \ldots, n - 1$ give the n distinct values.

10. Point sets

By the relation of absolute value to distance, the points z for which $|z| = r$ lie on a circle with center at the origin and radius r. Again, the points for which $|z| < r$ are the points interior to the circle. Similarly, $|z - z_0| = r$ is the equation of a circle of center z_0 and radius r, while the relation $|z - z_0| < r$ characterizes all points interior to this circle.

A *neighborhood*, or an *ϵ-neighborhood*, of a point z_0 is the set of points z such that $|z - z_0| < \epsilon$, where ϵ is a given positive number. A point z_0 is said to be a *limit point* of a set of points S if every neighborhood of z_0 contains a point of S distinct from z_0. This definition implies that every neighborhood of a limit point contains an infinite number of points of S. For the neighborhood $|z - z_0| < \epsilon_1$ contains a point z_1 of S distinct from z_0. Put $\epsilon_2 = \frac{1}{2}|z_1 - z_0| > 0$, and the neighborhood $|z - z_0| < \epsilon_2$ contains a point z_2 of S distinct from z_0. Put $\epsilon_3 = \frac{1}{2}|z_2 - z_0| > 0$, and continue indefinitely.

A limit point of S may or may not be a point of S. If every limit point belongs to the set, we say that the set is *closed*. There are two distinct types of limit points of a set, interior points and boundary points. A limit point z_0 is said to be an *interior point* of a set S if there exists a neighborhood of z_0 which consists entirely of points of S. A limit point of S which is not an interior point of S, that is, a point z_0 such that any neighborhood of z_0 contains both points of S and points which do not belong to S, is called a *boundary point* of S. If every point of a set is an interior point, we say that the set is *open*.

A set of points is *bounded* if there exists a constant M such that $|z| < M$ for all points of the set. The set is unbounded if no such number M exists.

A set of points S is said to be connected if any two of its points can be joined by a continuous curve all of whose points belong to S. An open connected set is called a *domain*, or *open* region. If we add to any set S those limit points which are not already in S, we obtain a closed set called the *closure* of S, often denoted by $\bar{S}$. In particular a set consisting of a domain together with its boundary points is called a *closed region*.

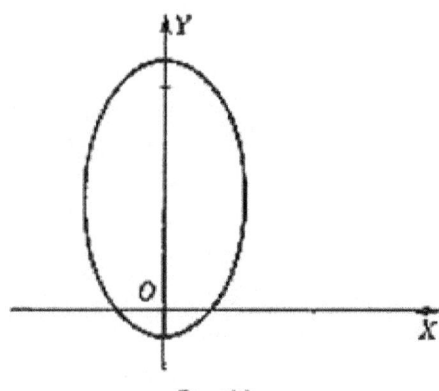

Fig. 10

Let us illustrate these terms. We note that the relation $|z| + |z - 8i| = 10$ is the equation of an ellipse with major axis 10, and foci at $(0,0)$ and $(0,8)$. See Fig. 10. Then the points inside the ellipse satisfy the relation

$$|z| + |z - 8i| < 10, \tag{55}$$

which defines a *domain*, or set which is *open* and *connected*. The relation

$$|z| + |z - 8i| \leq 10 \tag{56}$$

defines a closed region, the *closure* of the domain defined by Eq. (55). The set of z such that

either $\quad\quad |z| + |z - 8i| < 10,$

or $\quad\quad \operatorname{Re} z > 0 \quad \text{and} \quad |z| + |z - 8i| = 10, \tag{57}$

obtained from the domain of Eq. (55) by adding the right half of the boundary is neither open nor closed.

Since $|z| < 12$ for each of the three point sets defined by Eqs. (55) to (57), each of these sets is *bounded*. Each of these sets has as its limit points the points which make up the closed region of Eq. (56). Each of these sets has as its interior points the points which make up the domain of Eq. (55).

The points outside the ellipse satisfy the relation

$$|z| + |z - 8i| > 10. \qquad (58)$$

This defines an unbounded set, which is open and connected and so is a domain. The boundary points of each of the four point sets defined by Eqs. (55) to (58) are the points on the ellipse $|z| + |z - 8i| = 10$.

11. The Heine-Borel theorem

Let the points z of a set S be related to a set of circles C in the following way. For each point z we may find some circle of the set C, which we designate by C_z, such that z is an interior point of this circle. We may use any circle of the set C for several, or even for an infinite number, of its interior points. Again, it may be possible to find several, or even an infinite number of circles of the set C which will serve for a particular point z. We describe the relation of the set of circles C or C_z to S by saying that the set of circles C_z *covers* the points z of the set S.

The *Heine-Borel Covering theorem* states that:

Let an infinite set of circles C_z cover the points z of a set S. Then, if S is closed and bounded, a finite number of these circles C_z may be selected, such that this finite subset of circles covers the set S.

For a proof, we refer the reader to Chapter 1 of the author's *A Treatise on Advanced Calculus*.

Exercise 5

Each problem of this exercise contains a relation which describes a closed region. The equality sign, alone, gives the boundary points.

The inequality sign, alone, gives the domain made up of interior points.

Verify that each of Problems 1 to 5 represents a half-plane, or unbounded closed region bounded by a single straight line.

1. $\operatorname{Re} z \geq 0$.   2. $\operatorname{Im} z \leq 0$.   3. $\operatorname{Re}(z - iz) \geq 2$.
4. $|z - 1| \leq |z + 3|$.   5. $1 \leq \arg z \leq 1 + \pi$.

Verify that each of Problems 6 to 10 represents a bounded closed circular region.

6. $|z| \leq 2$.   7. $|1/z| \geq \frac{1}{2}$.   8. $|z - 1| \geq 4|z + 1|$.
9. $|2z + 3 + 5i| \leq 4$.   10. $z\bar{z} - (2 + i)z - (2 - i)\bar{z} \leq 4$.

Verify that Problems 11 and 12 represent circular rings.

11. $3 \leq |z| \leq 4$.   12. $1 \leq |z - i| \leq 2$.

Verify that Problems 13 and 14 represent bounded elliptical regions.

13. $|z - 2| + |z + 2| \leq 6$.   14. $|z - 4i| + |z + 4i| \leq 12$.

Verify the description of the following unbounded regions.

15. $|z| \geq 3$, the exterior of a circle.
16. $|1/z| \leq 2$, the exterior of a circle.
17. $|z - i| + |z + i| \geq 5$, the exterior of an ellipse.
18. $|z - 2| - |z + 2| \geq 1$, a hyperbolic region.
19. $|z - 2| - |z + 2| \leq 1$, a hyperbolic region.

2

FUNCTIONS OF A COMPLEX VARIABLE

Having discussed the algebra of complex numbers, we may now consider the differential calculus of complex variables. We first define a function of a complex variable, and explain how the notions of limit and continuity may be applied to such a function. Then we define differentiation and mention several elementary properties of the derivative which carry over from the calculus of real quantities. We next investigate the conditions that the differentiability of $f(z)$ imposes on the two real functions associated with it by the relation $f(z) = u(x,y) + iv(x,y)$. This leads us to the Cauchy-Riemann equations. Finally, we show that each of the functions u and v is a solution of Laplace's equation, a fact which is the basis of many applications of our subject to physical problems.

12. Functions

Let z denote the complex number represented by any point of S, an arbitrary point set in the complex plane. Then we call z a *complex variable* and S its range. We say that a second complex variable w

is a function of z on S if for each value of z in S, one or more values of w are determined. We write

$$w = f(z) \tag{1}$$

to indicate this relation. If there is just one value of w for each value of z, we say that w is a *single-valued* function of z.

Let $z = x + iy$ and $w = u + iv$. Then x and y determine z, which in turn determines w and hence u and v. Thus u and v are each functions of x and y, and

$$w = u(x,y) + iv(x,y). \tag{2}$$

13. Continuity

We often take the range of z to be a domain of simple type, such as the interior of a circle or rectangle. We call z a continuous variable whenever its range is D, a domain, or open region as defined in Section 10. For a continuous variable, if z_0 is a point of the range, z may take on every value in some neighborhood of z_0, or every value represented by a point sufficiently close to z_0.

Let $w = f(z)$ be a single-valued function defined in the domain D. Then for z_0 in D, the value $f(z_0)$ is determined. The function is continuous at z_0 if, in addition

$$\lim_{z \to z_0} f(z) \quad \text{exists and} \quad \lim_{z \to z_0} f(z) = f(z_0). \tag{3}$$

The meaning of this limiting relation is that, for any given positive quantity ϵ, it is always possible to assign a positive number $\delta(\epsilon)$, such that

$$|\Delta w| = |w - w_0| = |f(z) - f(z_0)| < \epsilon, \tag{4}$$

if

$$|\Delta z| = |z - z_0| < \delta. \tag{5}$$

Let us further suppose that the function $f(z)$ is defined for some of the boundary points of the domain D. Then by the continuity of $f(z)$ at the boundary point z_0 we mean that $f(z_0)$ is determined, and that the conditions for continuity are fulfilled at least if the z which appear in them lie within the domain. Thus in place of all z in the neighborhood $|z - z_0| < \delta$, we consider only all z which belong to the domain and for which $|z - z_0| < \delta$. This defines the phrase "continuity at a boundary point from the interior." Similarly for a function defined on a curved arc the phrase "continuity along the arc" means that the conditions for continuity are fulfilled for all points on the arc, without regard to the values of the function for other points.

Suppose that by making w_0 correspond to z_0, a boundary point of the range of definition of $f(z)$, we make $f(z)$ continuous at z_0 from the interior. Even though originally $f(z)$ was undefined at z_0, or defined to be different from w_0, we say that $f(z)$ assumes the boundary value w_0 at z_0.

The interpretation of $|\Delta z|$ as a distance $\sqrt{\Delta x^2 + \Delta y^2}$, and

$$|\Delta u| \leq |\Delta w|, \quad |\Delta v| \leq |\Delta w| \tag{6}$$

show that if $f(z)$ is continuous at $z_0 = x_0 + iy_0$, then each of the real functions $u(x,y)$ and $v(x,y)$ of Eq. (2) is continuous at (x_0, y_0). And the converse relation follows from the fact that

$$|\Delta w| \leq |\Delta u| + |\Delta v|. \tag{7}$$

In general, the number $\delta(\epsilon)$ of the relation (5) will depend on z_0 as well as on ϵ. However, if the range of definition is a *bounded closed region*, R, for each positive ϵ there exists a number δ *independent* of z_0 for which the relations (4) and (5) hold. That is, the function $f(z)$ is *uniformly continuous*. It is also true that a function $f(z)$ continuous in a bounded closed region R is necessarily bounded in R. That is, for z in R, a constant M exists for which

$$|f(z)| < M. \tag{8}$$

These properties follow from the corresponding properties of real functions of two real variables, since the continuity of $f(z)$ implies that of $u(x,y)$, $v(x,y)$, and we may apply Eq. (7) and the relation $|f(z)| = \sqrt{u^2 + v^2}$.

However, we shall illustrate the nature of the proof of this class of theorems by establishing the results without resolving $f(z)$ into its real and imaginary components. We first note that it follows from the property of continuity that if $f(z)$ is continuous at z_0, a number $\delta(\epsilon)$ exists such that

$$|f(z_2) - f(z_1)| < \epsilon \text{ if } |z_2 - z_0| < \delta \text{ and } |z_1 - z_0| < \delta. \tag{9}$$

In fact, if $|f(z) - f(z_0)| < \epsilon/2$ for $|z - z_0| < \delta$, Eq. (9) follows. The oscillation of $f(z)$ on any set is the least upper bound of $|f(z_2) - f(z_1)|$ for z_1 and z_2 each in the set. Thus the $\delta(\epsilon)$ of Eq. (9) is a number such that, in the circle with center z_0 and radius δ, the oscillation of $f(z)$ is less than ϵ. We may now prove the theorem:

If $f(z)$ is continuous in a closed and bounded region R, it is uniformly continuous in R.

For any given ϵ, and each z_0 in R, select the number δ_0 such that for z_1 and z_2 each in the circle $|z - z_0| < \delta_0$, $|f(z_2) - f(z_1)| < \epsilon$. Then take the circle with radius halved, $|z - z_0| < \delta_0/2$, as the circle C_s of the covering set for R, the $\overline{S}$ of Section 11. Then by the Heine-Borel covering theorem of Section 11, we may cover R with a finite subset of the circles C_s. Consider the least radius of any circle in the finite subset, and call it δ. Then for *any* two points z_1 and z_2 in R with $|z_2 - z_1| < \delta$, there is a circle of the subset, say $|z - z_3| < \delta_3/2$, which covers z_1, or includes z_1 in its interior. And since $\delta \leq \delta_3/2$, we have

$$|z_2 - z_3| \leq |z_2 - z_1| + |z_1 - z_3| < \delta + \delta_3/2 \leq \delta_3. \qquad (10)$$

Thus z_1 and z_2 are both interior to the circle $|z - z_3| < \delta_3$. And by the choice of δ_3, $|f(z_2) - f(z_1)| < \epsilon$. This proves the uniform continuity.

The boundedness of $f(z)$ follows as a corollary. Select a particular ϵ and let there be N circles in the finite covering subset C_f. Then if z_0 is a particular point in R, fixed throughout this argument, and z_t is any other point in R, we may join z_0 and z_t by a broken line with at most one chord in each of the circles C_f. It follows that $|f(z_t) - f(z_0)| \leq N\epsilon$, so that

$$|f(z_t)| < |f(z_0)| + N\epsilon + 1, \qquad (11)$$

and we may take this number as the M of Eq. (8).

14. Derivatives

Let $w = f(z)$ be a single-valued function defined in some domain or open region D. And let z_0 be a fixed complex number in D. Then when z is a complex variable, so is $\Delta z = z - z_0$. Moreover, as z tends to z_0, then Δz tends to zero. And in limits as $z \to z_0$, we need consider only values of z in D. Then $f'(z_0)$, the *derivative* of $w = f(z)$ with respect to z at z_0, is *defined* by

$$f'(z_0) = \lim_{z \to z_0} \frac{f(z) - f(z_0)}{z - z_0} = \lim_{\Delta z \to 0} \frac{f(z_0 + \Delta z) - f(z_0)}{\Delta z}, \qquad (12)$$

provided that the limit exists. In this case $w = f(z)$ is said to be *differentiable* at z_0.

A more explicit definition of $f'(z_0)$ is the number such that, for any positive ϵ, there is a δ such that

$$\left| \frac{f(z) - f(z_0)}{z - z_0} - f'(z_0) \right| < \epsilon \qquad (13)$$

for all values of z in D such that $0 < |z - z_0| < \delta$.

A single-valued function $f(z)$ is said to be *analytic at a point* z_0, if it has a derivative at z_0 and at all values in some neighborhood of z_0, that is in some two-dimensional region including z_0 as an interior point. We say that a function is analytic in a domain D if it is analytic at all points of D. (Some authors use the terms *regular*, *regular analytic*, or *holomorphic*, in place of analytic.) When $f(z)$ is analytic in D, then $f'(z)$ is a new function of z in D.

Let $\Delta w = f(z + \Delta z) - f(z)$. Then

$$f'(z) = \lim_{\Delta z \to 0} \frac{f(z + \Delta z) - f(z)}{\Delta z} = \lim_{\Delta z \to 0} \frac{\Delta w}{\Delta z} = \frac{dw}{dz}, \qquad (14)$$

where the last expression dw/dz is a suggestive alternative notation for $f'(z)$.

15. Differentiation rules

A number of elementary rules for differentiation may be proved from the definition of a derivative by the same arguments as those used for real variables. In the following, k is a constant, and F and G are differentiable functions of z. Then we have

$$\frac{d}{dz}(kF) = k\frac{dF}{dz}, \qquad (15)$$

$$\frac{d}{dz}(F + G) = \frac{dF}{dz} + \frac{dG}{dz}, \qquad (16)$$

$$\frac{d}{dz}(FG) = F\frac{dG}{dz} + G\frac{dF}{dz}, \tag{17}$$

$$\frac{d}{dz}\left(\frac{F}{G}\right) = \frac{G\,dF/dz - F\,dG/dz}{G^2}, \quad \text{if } G \neq 0. \tag{18}$$

Let $w = f(g)$ and $g = g(z)$. And for $g_0 = g(z_0)$, let $f(g)$ have a derivative at g_0 and $g(z)$ have a derivative at z_0. Then the composite function $w = f[g(z)] = F(z)$ has a derivative at z_0 given by the chain rule

$$\frac{dw}{dz} = \frac{dw}{dg}\frac{dg}{dz} \quad \text{or} \quad F'(z_0) = f'(g_0)g'(z_0). \tag{19}$$

This shows that an analytic function of an analytic function is again analytic.

It follows directly from the definition of a derivative that

$$\frac{dk}{dz} = 0 \quad \text{and} \quad \frac{dz}{dz} = 1, \tag{20}$$

and then by repeated application of Eq. (17) that for any positive integer n, $(d/dz)(z^n) = nz^{n-1}$. This formula may then be shown to hold for negative integers n by using Eq. (18).

Thus polynomials and rational functions of z may be differentiated by the familiar rules of the calculus of real variables. Hence any polynomial

$$P(z) = a_0 + a_1 z + a_2 z^2 + \ldots + a_n z^n \tag{21}$$

is analytic for every value of z. And any rational function, that is the ratio of two polynomials,

$$\frac{P(z)}{Q(z)} = \frac{a_0 + a_1 z + a_2 z^2 + \ldots + a_n z^n}{b_0 + b_1 z + b_2 z^2 + \ldots + b_m z^m} \tag{22}$$

is analytic for every value of z except the m roots of $Q(z) = 0$. If a function is analytic at *some* point in every neighborhood of z_0, but is not analytic at z_0 itself, then z_0 is called a *singular point* or a singularity of the function. Thus any z_0 which makes $P(z_0) \neq 0$ but $Q(z_0) = 0$ is a singular point of the rational function $P(z)/Q(z)$.

Exercise 6

Using the notation $F(z + \Delta z) = F + \Delta F$, $G(z + \Delta z) = G + \Delta G$, and Eq. (14), give a proof of

1. Eq. (15). 2. Eq. (16). 3. Eq. (17).

Use the rules of Section 15 to verify that

4. If $w = 2z^3 - 3z + 5$, then $dw/dz = 6z^2 - 3$.
5. If $w = (z^2 + 1)^4$, then $dw/dz = 8z(z^2 + 1)^3$.
6. If $w = \dfrac{z + 2}{z - 2}$, then $\dfrac{dw}{dz} = \dfrac{-4}{(z - 2)^2}$ for $z \neq 2$.

Let us write Δz in polar form as $r \operatorname{cis} \theta$. Then from Sections 7 and 8, $\Delta \bar{z} = \overline{\Delta z} = r \operatorname{cis}(-\theta)$. Deduce from this that

7. $\Delta \bar{z}/\Delta z = \operatorname{cis}(-2\theta) = \cos 2\theta - i \sin 2\theta$, so that $w = \bar{z}$ is nowhere differentiable.
8. $w = x = \operatorname{Re} z = (z + \bar{z})/2$ is nowhere differentiable.
9. $w = y = \operatorname{Im} z = (z - \bar{z})/2i$ is nowhere differentiable.
10. If $w = |z|^2 = z\bar{z}$, then $\Delta w/\Delta z = \bar{z} + \Delta \bar{z} + z \operatorname{cis}(-2\theta)$, so that at $z = 0$, $dw/dz = 0$, but for $z \neq 0$, w is not differentiable.
11. Test the function $w = x^3 y(x + iy)/(x^4 + y^2)$ when $z \neq 0$, $w(0) = 0$, for differentiability at $z = 0$. Note that $\Delta w/\Delta z = x^2 y/(x^4 + y^2)$, which approaches zero when $\Delta z = z$ approaches zero along $x = 0$, or any line $y = mx$, but if $y = kx^2$, the limit is $k/(1 + k^2)$ which varies with k, so that the limit of Eq. (12) or (14) does not exist, and there is no derivative at $z = 0$.

16. Cauchy-Riemann equations

If the function $w = f(z)$ has a derivative $f'(z)$, then by Eq. (14) we have

$$f'(z) = \frac{dw}{dz} = \lim_{\Delta z \to 0} \frac{\Delta w}{\Delta z}. \qquad (23)$$

Let us investigate the conditions that differentiability imposes on the real functions $u(x,y)$ and $v(x,y)$ where, as in Eq. (2), $w = u + iv$. The relation $w = u + iv$ implies that

$$\Delta w = \Delta u + i\, \Delta v \quad \text{and} \quad \frac{\Delta w}{\Delta z} = \frac{\Delta u}{\Delta z} + i\frac{\Delta v}{\Delta z} \qquad (24)$$

Sec. 16   FUNCTIONS OF A COMPLEX VARIABLE   27

The limit in Eq. (23) necessarily exists for any sequence of values of $\Delta z = \Delta x + i\,\Delta y$ which approach zero. In particular, if we take $\Delta y = 0$, and let $\Delta z = \Delta x$ approach zero, we find from Eqs. (23) and (24) that

$$f'(z) = \lim_{\Delta x \to 0}\left(\frac{\Delta u}{\Delta x} + i\frac{\Delta v}{\Delta x}\right) = \frac{\partial u}{\partial x} + i\frac{\partial v}{\partial x} = u_x + iv_x. \qquad (25)$$

Again, if we take $\Delta x = 0$, and let $\Delta z = i\,\Delta y$ approach zero, we find from Eqs. (23) and (24) that

$$f'(z) = \lim_{\Delta y \to 0}\left(\frac{1}{i}\frac{\Delta u}{\Delta y} + \frac{\Delta v}{\Delta y}\right) = -i\frac{\partial u}{\partial y} + \frac{\partial v}{\partial y} = -iu_y + v_y. \qquad (26)$$

From Eqs. (25) and (26) it follows that

$$f'(z) = u_x + iv_x = v_y - iu_y. \qquad (27)$$

We express our first conclusion from Eq. (27) as

Theorem 1. If the function $w = f(z) = u + iv$ is differentiable at $z = x + iy$, then each of the four partial derivatives $\partial u/\partial x = u_x$, $\partial u/\partial y = u_y$, $\partial v/\partial x = v_x$, $\partial v/\partial y = v_y$ exists at (x,y).

We next prove a theorem which we shall need when we discuss the complex integral calculus.

Theorem 2. If a function $w = f(z)$ is differentiable at each point of a connected domain D, and if its derivative $f'(z) = 0$ at each point of D, then $f(z) = c$, a complex constant, in D.

The hypothesis of the theorem that $f'(z) = 0$, together with Eq. (27), implies that u_x, u_y as well as v_x, v_y are each zero. Hence, by a theorem on real functions of two real variables with both partial derivatives zero in a connected domain, $u = c_1$ and $v = c_2$, so that $w = c_1 + ic_2 = c$, as the theorem asserts.

Theorem 3. If the function $w = f(z) = u + iv$ is differentiable at the point $z = x + iy$, then the relations

$$\frac{\partial u}{\partial x} = \frac{\partial v}{\partial y}, \qquad \frac{\partial u}{\partial y} = -\frac{\partial v}{\partial x}, \qquad (28)$$

must hold at (x,y).

Since u and v are real, their partial derivatives are real. Hence the second equality of Eq. (27) implies that

$$u_x = v_y, \quad v_x = -u_y, \qquad (29)$$

which is equivalent to Eq. (28). The relations expressed by Eq. (28) or (29) are known as the *Cauchy-Riemann* differential equations.

Theorems 1 and 3 express necessary conditions for the differentiability of $f(z) = u + iv$. *Sufficient* conditions are given by our next theorem.

Theorem 4. If $u(x,y)$ and $v(x,y)$ have partial derivatives which are continuous at (x,y) and satisfy the Cauchy-Riemann differential equations (28) for (x,y) then the function $f(z) = w = u + iv$ has a derivative $f'(z)$ at $z = x + iy$.

The assumed continuity of u_x and u_y at (x,y) implies that these partial derivatives exist in some neighborhood of (x,y). Hence for $h = \Delta x$ and $k = \Delta y$ sufficiently small, $(x + h, y + k)$ will be in this neighborhood, and we may write

$$\Delta u = u(x + h, y + k) - u(x,y)$$
$$= u(x + h, y + k) - u(x + h, y) + u(x + h, y) - u(x,y)$$
$$= k u_y(x + h, y + \theta_1 k) + h u_x(x + \theta_2 h, y), \qquad (30)$$

where $0 < \theta_1 < 1, 0 < \theta_2 < 1$, by the mean value theorem for functions of one real variable. From this and the continuity of u_x and u_y at (x,y), we have

$$\Delta u = h[u_x(x,y) + \epsilon_1] + k[u_y(x,y) + \epsilon_2], \qquad (31)$$

where ϵ_1 and ϵ_2 both approach zero with h and k. Similarly

$$\Delta v = h[v_x(x,y) + \epsilon_3] + k[v_y(x,y) + \epsilon_4], \qquad (32)$$

where ϵ_3 and ϵ_4 both approach zero with h and k.

From Eqs. (31) and (32) we obtain

$$\Delta w = \Delta u + i \Delta v = h(u_x + iv_x) + k(u_y + iv_y) + \omega_1 h + \omega_2 k \qquad (33)$$

where

$$\omega_1 = \epsilon_1 + i\epsilon_3, \quad \omega_2 = \epsilon_2 + i\epsilon_4. \qquad (34)$$

Since Eq. (28) holds at (x,y), in Eq. (33) we may replace u_y by $-v_x$ or $i^2 v_x$, and v_y by u_x, and so deduce that

$$\Delta w = (h + ik)(u_x + iv_x) + \omega_1 h + \omega_2 k. \qquad (35)$$

But $h + ik = \Delta x + i \Delta y = \Delta z$, so that

$$\frac{\Delta w}{\Delta z} = u_x + iv_x + \omega_1 \frac{h}{\Delta z} + \omega_2 \frac{k}{\Delta z}. \qquad (36)$$

Sec. 16   FUNCTIONS OF A COMPLEX VARIABLE   29

Again $|\Delta z| = \sqrt{h^2 + k^2}$, $|\Delta z| \geq h$, $|\Delta z| \geq k$ implies that

$$\left|\frac{h}{\Delta z}\right| \leq 1, \quad \left|\frac{k}{\Delta z}\right| \leq 1 \tag{37}$$

and also shows that $h \to 0$, $k \to 0$ when $\Delta z \to 0$. But when $h \to 0$, $k \to 0$ so do the ϵ's, and hence the ω's, by Eq. (34). It follows from Eq. (37) that the last two terms in Eq. (36) tend to zero with Δz. Consequently we have

$$f'(z) = \lim_{\Delta z \to 0} \frac{\Delta w}{\Delta z} = u_x + iv_x. \tag{38}$$

Thus the function $f(z) = w = u + iv$ has a derivative, as was to be proved.

Exercise 7

Verify that the Cauchy-Riemann differential equations, and the relation of Eq. (27), holds for each of the following functions.

1. $u = 2x - 3y$, $v = 3x + 2y$, $w = (2 + 3i)z$, $dw/dz = 2 + 3i$.
2. $u = -2xy$, $v = x^2 - y^2$, $w = iz^2$, $dw/dz = 2iz$.
3. $u = x^3 - 3xy^2$, $v = 3x^2y - y^3$, $w = z^3$, $dw/dz = 3z^2$.
4. $u = -x/(x^2 + y^2)$, $v = y/(x^2 + y^2)$, $w = -1/z$, $dw/dz = 1/z^2$, $z \neq 0$.

5. Verify that if $z = r \operatorname{cis} \theta$, $\partial z/\partial r = \operatorname{cis} \theta$, $\partial z/\partial \theta = ir \operatorname{cis} \theta$. Let $f(z)$ have a derivative. By taking $\Delta \theta = 0$, deduce that $u_r + iv_r = f'(z) \partial z/\partial r = f'(z) \operatorname{cis} \theta$. Then by taking $\Delta r = 0$, deduce that $u_\theta + iv_\theta = f'(z) \partial z/\partial \theta = f'(z) ir \operatorname{cis} \theta$. Hence show that

$$f'(z) = (u_r + iv_r) \operatorname{cis}(-\theta) = -\frac{i}{r}(u_\theta + iv_\theta) \operatorname{cis}(-\theta).$$

Also show that

$$\frac{\partial u}{\partial r} = \frac{1}{r}\frac{\partial v}{\partial \theta}, \quad \frac{\partial v}{\partial r} = -\frac{1}{r}\frac{\partial u}{\partial \theta},$$

which are the Cauchy-Riemann differential equations in polar form.

6. Use the transformation $x = r \cos \theta$, $y = r \sin \theta$ to show that the results of Problem 5 are equivalent to Eqs. (27) and (28).

Verify that the results of Problem 5 hold for each of the following functions.

7. $u = 2r \sin \theta$, $v = -2r \cos \theta$, $w = -2iz$, $dw/dz = -2i$.

8. $u = -\dfrac{\cos\theta}{r}$, $v = \dfrac{\sin\theta}{r}$, $w = -\dfrac{1}{z}$, $\dfrac{dw}{dz} = \dfrac{1}{z^2}$, $z \neq 0$.

9. $u = r^n \cos n\theta$, $v = r^n \sin n\theta$, $w = z^n$, $dw/dz = nz^{n-1}$.

Use the Cauchy-Riemann test to prove that each of the following functions is nowhere differentiable.

10. $w = x$.   11. $w = y$.   12. $w = 3ix$.   13. $w = ie^y$.

Prove that a function which is differentiable in a region must be a constant there if throughout the region

14. $u = 0$.   15. $v = 0$.   16. $\overline{f(z)}$ is differentiable.

17. $|f(z)|$ is constant. *Hint*: From $u^2 + v^2 = a$, $uu_x + vv_x = 0$, $uu_y + vv_y = -uv_x + vu_x = 0$; $(u^2+v^2)u_x = 0$, $(u^2+v^2)v_x = 0$.

Prove that each of the following functions is differentiable only at the points indicated, and hence is nowhere analytic.

18. $w = x^2$, $(0,y)$.   19. $w = y^2$, $(x,0)$.   20. $w = |z|^2$, $(0,0)$.
21. $w = \cos x$, $(n\pi, y)$.   22. $w = i \sin y$, $(x, n\pi + \pi/2)$.

23. Show that the function $w = \sqrt{|xy|}$ is not differentiable at the origin, but that the Cauchy-Riemann equations are satisfied at the origin. *Hint*: At $z = 0$, $\Delta z = z$. If $y = k^2 x$, as $x \to 0$ through positive values, $\Delta w/\Delta z = k/(1 + ik^2)$.

24. By putting $x = \tfrac{1}{2}(\bar{z} + z)$, $y = (i/2)(\bar{z} - z)$, we may regard $w = u(x,y) + iv(x,y)$ formally as a function of two independent (complex) variables z and $\bar{z}$. Show that

$$\frac{\partial w}{\partial \bar{z}} = \frac{\partial w}{\partial x}\frac{\partial x}{\partial \bar{z}} + \frac{\partial w}{\partial y}\frac{\partial y}{\partial \bar{z}} = (u_x + iv_x)\tfrac{1}{2} + (u_y + iv_y)(i/2) = 0$$

leads to Eq. (28). And that in view of this, $f'(z) = \partial w/\partial z$ is equivalent to Eq. (27).

25. Use Problem 24 to check Problems 10 to 12 by showing that $\partial w/\partial \bar{z} \neq 0$ if $w = \tfrac{1}{2}(z + \bar{z})$, $w = (i/2)(\bar{z} - z)$, or $w = (3i/2)(z + \bar{z})$.

17. Laplace's equation

Let the function $f(z) = w = u + iv$ be analytic throughout a domain D. Then by definition $f(z)$ is differentiable throughout D. By Theorem 1 of Section 16, this implies the existence of the first partial derivatives u_x, u_y, v_x, v_y. It will be proved in Section 53 that the existence of $f'(z)$ in D implies the existence and continuity of the partial derivatives of u and v of all higher orders. Let us anticipate this result to the extent of assuming the existence and continuity of the second partial derivatives. In particular, this implies that

$$u_{xy} = u_{yx} \quad \text{and} \quad v_{xy} = v_{yx}. \tag{39}$$

By Theorem 3 of Section 16, the Cauchy-Riemann equations (28) must hold for any (x,y) in D, so that

$$u_x = v_y, \qquad u_y = -v_x. \tag{40}$$

It follows by differentiation of these, and use of Eq. (39) that

$$u_{xx} = v_{yx}, \quad u_{yy} = -v_{xy} = -v_{yx} \quad \text{and hence } u_{xx} + u_{yy} = 0. \tag{41}$$

$$v_{yy} = u_{xy} = u_{yx}, \quad v_{xx} = -u_{yx} \quad \text{and hence } v_{xx} + v_{yy} = 0. \tag{42}$$

Thus each of the functions u and v must be a solution of *Laplace's equation* in two dimensions:

$$\nabla^2 \phi = \frac{\partial^2 \phi}{\partial x^2} + \frac{\partial^2 \phi}{\partial y^2} = 0. \tag{43}$$

This equation is of fundamental importance in many branches of applied mathematics. Among the functions of physical significance which satisfy it are the following: the potential at a point interior to a domain devoid of matter in a gravitational field, the potential at a point interior to a domain devoid of charges in an electric field, the velocity potential and stream function of two-dimensional irrotational flow of an incompressible nonviscous fluid.

Any function $\phi(x,y)$ is said to be *harmonic* in a domain D if its first and second partial derivatives exist and are continuous in D, and if Laplace's Eq. (43) holds throughout D. A function is harmonic at a point (x,y) if it is harmonic in some neighborhood of (x,y). Our discussion shows that from any analytic function of z, $f(z) = u + iv$, we may obtain two harmonic functions u and v. This fact is the basis for many applications of the theory of functions of a complex variable to the determination of two-dimensional potential functions.

If $u + iv = f(z)$ is an analytic function of z, then u and v are called *conjugate* (harmonic) *functions*. Note that neither u nor v can be chosen arbitrarily. In fact, each alone must be a harmonic function, and the two together must satisfy the Cauchy-Riemann equations.

Several properties of such functions are given in Problems 1, 5, 6, 11, 15-17, and 23 of Exercise 8.

Exercise 8

1. Let $u(x,y)$ be a single-valued harmonic function in D, the circular domain $x^2 + y^2 < a^2$. Show that $-u_y\, dx + u_x\, dy$ is an exact differential, and so defines to within an additive constant a function $v(x,y)$ such that $dv = -u_y\, dx + u_x\, dy$. From $dv = v_x\, dx + v_y\, dy$, deduce that Eq. (40) holds, which proves that u and v are conjugate functions in D.

Verify that each given u is harmonic. Then use the result of Problem 1, $dv = -u_y dx + u_x dy$, to show that

2. If $u = 2x - 3y$, then $v = 3x + 2y + C$.
3. If $u = -2xy$, then $v = x^2 - y^2 + C$.
4. If $u = x^3 - 3xy^2$, then $v = 3x^2y - y^3 + C$.

5. If u,v are conjugate harmonic functions, show that $v,-u$ are conjugate harmonic functions. *Hint:* $v - iu = -i(u + iv)$.

6. Let $v(x,y)$ be a single-valued harmonic function in D, the circular domain $x^2 + y^2 < a^2$. By the argument of Problem 1, show that $du = v_y\, dx - v_x\, dy$ defines to within an additive constant a function $u(x,y)$ such that u,v are conjugate harmonic functions.

7. Use Problems 5 and 1 to check Problem 6.

Verify that each given v is harmonic. Then use the result of Problem 6, $du = v_y\, dx - v_x\, dy$, to show that

8. If $v = x^2 - y^2$, then $u = -2xy + C$.
9. If $v = x^3 - 3xy^2$, then $u = -3x^2y + y^3 + C$.

Let u and v be conjugate functions, and $f(z) = u + iv$. Prove

10. $u_x v_x + u_y v_y = 0$, or $(-u_x/u_y)(-v_x/v_y) = -1$, if $u_y v_y \neq 0$.

11. The families of curves $u(x,y) = c_1$ and $v(x,y) = c_2$ are *orthogonal*, or intersect at right angles, at any point of intersection where $f'(z) \neq 0$. *Hint:* From $u_x\, dx + u_y\, dy = 0$, $m_1 = -u_x/u_y$. From $v_x\, dx + v_y\, dy = 0$, $m_2 = -v_x/v_y$. By Problem 10, $m_1 m_2 = -1$ if $u_y v_y \neq 0$. Since $f'(z) \neq 0$, $u_y = -v_x$ and $u_x = v_y$ are not both zero. If either one is zero, one curve is parallel to OX and the other to OY.

Illustrate Problem 11 by plotting a few curves of each family for

12. $f(z) = (1 - 2i)z = x + 2y + i(-2x + y)$, straight lines.

13. $f(z) = \dfrac{1}{z} = \dfrac{x}{x^2 + y^2} - i\dfrac{y}{x^2 + y^2}$, circles.

14. $f(z) = z^2 = x^2 - y^2 + 2ixy$, $z \neq 0$, rectangular hyperbolas.

15. Show that the function $\log |f(z)| = \tfrac{1}{2} \log (u^2 + v^2)$ is analytic at any point (x,y) at which $f(z)$ is harmonic and $f(z) \neq 0$.

16. The polar form of the Cauchy-Riemann equations is $ru_r = v_\theta$, $rv_r = -u_\theta$, by Problems 5, 6 of Exercise 7. Deduce from these that u and v each satisfy

$$\phi_{rr} + \frac{1}{r}\phi_r + \frac{1}{r^2}\phi_{\theta\theta} = 0.$$

17. Use the transformation $x = r \cos \theta$, $y = r \sin \theta$ to show that $r\phi_r = x\phi_x + y\phi_y$ and $\phi_\theta = -y\phi_x + x\phi_y$. Also show that

$$r(r\phi_r)_r + \phi_{\theta\theta} = r^2(\phi_{xx} + \phi_{yy}),$$

which proves that the last equation of Problem 16 is equivalent to Laplace's equation.

Use the results of Problems 16 and 17 to verify that each of the following functions is harmonic.

18. $\phi = r^n \sin n\theta$.   19. $\phi = r^n \cos n\theta$.
20. $\phi = \log r$.   21. $\phi = \theta$.

22. $\phi = u\left(\dfrac{x}{x^2 + y^2}, \dfrac{y}{x^2 + y^2}\right)$ is harmonic if $u(x,y)$ is. *Hint:* If $\theta = \theta$, $r = 1/R$, then $R(R\phi_R)_R = r(r\phi_r)_r$, and $x/(x^2 + y^2) = r \cos \theta/r^2 = R \cos \theta$.

23. *Method of Milne-Thomson.* As in Problem 24 of Ex. 7, let $x = \tfrac{1}{2}(z + \bar{z})$, $y = (i/2)(\bar{z} - z)$. Then, formally, $f(z) = u\left[\dfrac{z + \bar{z}}{2}, \dfrac{i(\bar{z} - z)}{2}\right] + iv\left[\dfrac{z + \bar{z}}{2}, \dfrac{i(\bar{z} - z)}{2}\right]$ is an identity in z and $\bar{z}$. For $\bar{z} = z$ it reduces to $f(z) = u(z,0) + iv(z,0)$. Similarly, from $f'(z) = u_x + iv_x = v_y - iu_y$, we may derive the relations $f'(z) = u_x(z,0) - iu_y(z,0) = v_y(z,0) + iv_x(z,0)$. When u or v is a given harmonic function these equations may be used to determine $f(z)$ by an integration.

Use the relation $f(z) = \int [u_x(z,0) - iu_y(z,0)] dz$ to show that for the given *harmonic* function u,

24. If $u = 2x - 3y$, $f(z) = \int (2 + 3i) dz = (2 + 3i)z + Ci$.
25. If $u = y^3 - 3x^2y$, $f(z) = \int 3iz^2 dz = iz^3 + Ci$.

Use the relation $f(z) = \int [v_y(z,0) + iv_x(z,0)] dz$ to show that for the given *harmonic* function v,

26. If $v = x^2 - y^2$, then $f(z) = \int 2iz \, dz = iz^2 + C$.
27. If $v = x/(x^2 + y^2)$, then $f(z) = \int -i \, dz/z^2 = i/z + C$.

3

POWER SERIES

In this chapter we discuss infinite series whose terms are complex constants or complex functions. We are particularly interested in power series, and shall show that every convergent power series represents a differentiable function. Because of this property, we shall find power series a convenient means of defining the exponential and certain other elementary functions for complex values of the variable in Chapter 4.

18. Infinite series

Any infinite sequence of complex numbers z_n determines an *infinite series*:

$$\sum_{n=1}^{\infty} z_n = z_1 + z_2 + \ldots + z_n + \ldots. \tag{1}$$

If $z_n = x_n + iy_n$, the nth partial sum of the series is

$$Z_N = \sum_{n=1}^{N} z_n = \sum_{n=1}^{N} (x_n + iy_n) = \sum_{n=1}^{N} x_n + i \sum_{n=1}^{N} y_n$$
$$= X_N + iY_N. \tag{2}$$

The series (1) is said to be *convergent* if

$$\lim_{N \to \infty} Z_N = Z = X + iY, \quad \text{or if} \quad \lim_{N \to \infty} |Z_N - Z| = 0. \tag{3}$$

But $|Z_N - Z| < \epsilon$ if and only if

$$|Z_N - Z|^2 = (X_N - X)^2 + (Y_N - Y)^2 < \epsilon^2, \tag{4}$$

so that the relation of Eq. (3) is equivalent to

$$\lim_{N \to \infty} X_N = X, \quad \lim_{N \to \infty} Y_N = Y. \tag{5}$$

Thus if the real series Σx_n converges to the sum X, and the real series Σy_n converges to the sum Y, then the complex series Σz_n converges to the sum $Z = X + iY$. If either Σx_n or Σy_n is divergent, then Σz_n is divergent also.

If a real series is convergent, its nth term must approach zero when n becomes infinite. Consequently the convergence of Σz_n, which requires the convergence of Σx_n and Σy_n, implies that

$$\lim_{n \to \infty} x_n = 0, \quad \lim_{n \to \infty} y_n = 0. \tag{6}$$

Since $z_n = x_n + iy_n$, it follows that

$$\lim_{n \to \infty} |z_n| = \lim_{n \to \infty} \sqrt{x_n^2 + y_n^2} = 0. \tag{7}$$

Since this holds, we can choose an integer N_1 such that for $n > N_1$, $|z_n| < 1$. But the terms with $n \leq N_1$ are finite in number. Hence we can choose a number M_1 such that $|z_n| < M_1$ for $n \leq N_1$. Hence, if $M = 1 + M_1$, we shall have

$$|z_n| < M \tag{8}$$

for all n. We have thus proved the following results:

If a series of complex terms (1) converges, its nth term must approach zero, Eq. (7). And its terms are all uniformly bounded in absolute value, Eq. (8).

Let us next assume that the series $\Sigma |z_n|$ converges. We have

$$|z_n| = \sqrt{x_n^2 + y_n^2}, \quad |x_n| \leq |z_n|, \quad \text{and} \quad |y_n| \leq |z_n|. \tag{9}$$

Hence, by the comparison test for series with positive real terms, each of the series $\Sigma |x_n|$ and $\Sigma |y_n|$ is convergent. But an absolutely convergent series of real terms is necessarily convergent, so that

Σx_n and Σy_n each converge. Thus the complex series Σz_n converges. We have thus proved that

If $\Sigma |z_n|$ is convergent, then Σz_n is convergent.

When $\Sigma |z_n|$ converges, the series Σz_n is said to *converge absolutely*. Thus it is also true for complex series that absolute convergence implies convergence.

19. Power series

A *power series* in z is an infinite series whose terms are products of integral powers of z by complex constants

$$\sum_{n=0}^{\infty} a_n z^n. \tag{10}$$

A first convergence property of such series is given by:

Theorem 1. If a power series in z converges for some value of z, say $z_1 \neq 0$, then it converges absolutely for all values of z with $|z| < |z_1|$.

By hypothesis, $\Sigma a_n z_1^n$ converges. Hence by the property derived from Eq. (8), there is a number M for which

$$|a_n z_1^n| < M. \tag{11}$$

Select any z with $0 < |z| < |z_1|$. And let us put

$$\frac{|z|}{|z_1|} = c, \quad \text{where } c < 1, \text{ since } |z| < |z_1|. \tag{12}$$

Hence, for the value of z under consideration, we have

$$|a_n z^n| = |a_n z_1^n| \left|\frac{z}{z_1}\right|^n < Mc^n. \tag{13}$$

The series $\Sigma |a_n z^n|$ and ΣMc^n each have positive real terms. And the second series ΣMc^n converges since it is a geometric series with ratio $c < 1$. Hence we may conclude from Eq. (13) and the comparison test for positive real series that the first series $\Sigma |a_n z^n|$ converges. Thus $\Sigma a_n z^n$ converges absolutely, as was to be proved.

Suppose that a power series converges for $z_1 \neq 0$, where $|z_1| = r_1$, and diverges for z_2, where $|z_2| = r_2$. Then by Theorem 1, the series will converge for all z with $|z| < r_1$. Similarly, it will diverge for all z

with $|z| > r_2$, since the convergence for any such value would imply that for z_2, contrary to our assumption. This leads to a separation of the positive values of r into two classes. One class contains values of r_1 such that the series converges for all z with $|z| = r_1$. The other class contains values of r_2 such that the series diverges for some z with $|z| = r_2$. Each positive value is either an r_1 or an r_2, every r_1 precedes every r_2, and there are some numbers in each class. Hence by Dedekind's theorem on real numbers there is a number R which effects the separation, in the sense that every $r_1 \leq R$ and every $r_2 \geq R$. The circle $|z| = R$ is called the *circle of convergence* of the power series. And the number R is called the *radius of convergence* of the power series. It may be defined as the number such that the series converges for all z with $|z| < R$, and diverges for all z with $|z| > R$. On the circle $|z| = R$, the series may converge everywhere, nowhere, or somewhere as illustrated in Problems 27-29 of Exercise 9.

Whenever our assumption that there is some $z_1 \neq 0$ and some z_2 holds, there is a value of R finite and different from zero. But these assumptions do not hold for all power series. If the power series converges for no values except $z = 0$, we put $R = 0$. And if the series converges for all values of z, we put $R = \infty$. In any case, R may be determined by the *Cauchy-Hadamard rule*, which we state as:

Theorem 2. Let a_n be the general coefficient of the power series $\Sigma a_n z^n$. Then the radius of convergence of the series is given by

$$R = \frac{1}{A}, \quad \text{where} \quad A = \varlimsup_{n \to \infty} \sqrt[n]{|a_n|}. \tag{14}$$

To establish this, we note that for any fixed value of z,

$$\varlimsup_{n \to \infty} \sqrt[n]{|a_n z^n|} = A|z|. \tag{15}$$

First suppose that $|z| > 1/A$. Then $A|z| > 1$. Hence by the definition of the superior limit $\varlimsup$ and Eq. (15), we must have $\sqrt[n]{|a_n z^n|} > 1$, or $|a_n z^n| > 1$, for an infinite number of subscripts n. This shows that $\Sigma a_n z^n$ diverges, since $a_n z^n$ cannot tend to zero as required by the necessary condition for convergence derived from Eq. (7).

Next suppose that $|z| < 1/A$. Then $A|z| < 1$. Hence by the definition of the superior limit $\overline{\lim}$ and Eq. (15), it must be possible to choose an integer N_1 such that $\sqrt[n]{|a_n z^n|} < t$ for all $n > N_1$, where $t < 1$. Consequently, for $n > N_1$, $|a_n z^n| < t^n$, but Σt^n is a geometric series with ratio $t < 1$, and so converges. Hence by the comparison test for positive real series, $\Sigma |a_n z^n|$ converges, and $\Sigma a_n z^n$ converges absolutely. This proves Theorem 2.

A specialized condition is

If $$\lim_{n \to \infty} \sqrt[n]{|a_n|} = L, \quad \text{then} \quad R = \frac{1}{L}. \tag{16}$$

And a more specialized condition is

If $$\lim_{n \to \infty} \left|\frac{a_{n+1}}{a_n}\right| = L, \quad \text{then} \quad R = \frac{1}{L} \tag{17}$$

This follows by a direct application of the test-ratio test to the series $\Sigma a_n z^n$. If the limit of Eq. (17) exists, it may be simpler to find than that of Eq. (16).

Exercise 9

Let R denote the radius of convergence of the power series $\sum_{n=1}^{\infty} a_n z^n$. Show that $R = 1$ if

1. $a_n = 1$.
2. $a_n = n$
3. $a_n = 1/n$.
4. $a_n = n(n+1)$.
5. $a_n = n^{\log n}$.
6. $a_n = 1/n^2$.
7. $a_{2m} = 1/3^m$, $a_{2m+1} = \sqrt{m}$.
8. $a_{2m} = m^2$, $a_{2m+1} = 1/4^m$.
9. $|a_n| = 25$.
10. $|a_n| = n$.
11. $|a_n| = n^2$.

Show that $R = e$ if

12. $a_n = (1 + 1/n)^{-n^2}$.
13. $a_n = (1 - 1/n)^{n^2}$.

Show that $R = 0$ if

14. $a_n = (\log n)^n$.
15. $a_n = n^n$.
16. $a_{2m} = (\log m)^m$, a_{2m+1} arbitrary.

Show that $R = \infty$ if

17. $a_n = 1/n^n$.
18. $a_n = 1/(\log n)^n$.

Use the test ratio directly, or Eq. (17), to show that if

19. $a_n = 1/n!$, $R = \infty$.
20. $a_n = n!$, $R = 0$.

21. $a_n = n!/n^n$, $R = e$, and deduce that $\lim_{n\to\infty} \sqrt[n]{n!}/n = 1/e$.

Given that $\overline{\lim}_{n\to\infty} \sqrt[n]{b_n} = B$, finite and not zero, show that if

22. $a_n = n^n b_n$, $R = 0$.   **23.** $a_n = b_n/n^n$, $R = \infty$.

24. $a_n = \alpha^n b_n$, where α is any complex number $\neq 0$, $R = 1/|\alpha|B$.

25. $a_n = n! b_n$, $R = 0$. *Hint:* Use the limit found in Problem 21.

26. $a_n = b_n/n!$, $R = \infty$. *Hint:* Use the limit found in Problem 21.

27. Show that the series of Problem 1, Σz^n with $R = 1$, *diverges* for *all* z with $|z| = 1$, on the circle of convergence.

28. Show that the series of Problem 6, $\Sigma z/n^2$ with $R = 1$, *converges* for *all* z with $|z| = 1$, on the circle of convergence.

29. Show that the series of Problem 3, $\Sigma z^n/n$ with $R = 1$, diverges for $z = 1$, but converges for $z = -1$.

20. Uniform convergence

Suppose that each of the functions $g_n(z)$ is defined for z in a point set S and the series

$$\sum_{n=1}^{\infty} g_n(z) = g_1(z) + g_2(z) + \ldots + g_n(z) + \ldots \qquad (18)$$

converges at every point in S. Then for any point z in S we may apply Eqs. (1) and (3) of Section 18 with $z_n = g_n(z)$. Write

$$\sum_{n=1}^{\infty} g_n(z) = G(z), \qquad \sum_{n=1}^{N} g_n(z) = G_N(z). \qquad (19)$$

Then $R_N(z)$, the *remainder after N terms*, is

$$R_N(z) = G(z) - G_N(z) = \sum_{n=N+1}^{\infty} g_n(z). \qquad (20)$$

And we may conclude that

$$|G(z) - G_N(z)| \to 0, \quad \text{or} \quad |R_N(z)| \to 0, \quad \text{as } N \to \infty. \qquad (21)$$

Hence for any given (arbitrarily small) positive ϵ, there exists an integer, and hence a *least* integer, $n_s(\epsilon)$ such that $|R_N(z)| < \epsilon$, or

$$|R_N(z)| = |g_{N+1}(z) + g_{N+2}(z) + \ldots| < \epsilon, \quad \text{if } N > n_s(\epsilon). \qquad (22)$$

The behavior of this smallest permissible $n_s(\epsilon)$ at a point z for a sequence of values of ϵ, such as $\epsilon_k = 1/2^k$, which approaches zero, is a measure of the rapidity of the convergence. Where the $n_z(\epsilon_k)$ are relatively large, the series converges slowly. Where the $n_z(\epsilon_k)$ are relatively small, the series converges rapidly.

For a fixed ϵ, we may consider a smallest permissible $n_z(\epsilon)$ for each z in S. Suppose that these numbers are bounded as a set. Then there exists an $n_1(\epsilon)$ which exceeds every $n_z(\epsilon)$ for z in S. It then follows that

$$|R_N(z)| = |g_{N+1}(z) + g_{N+2}(z) + \ldots| < \epsilon, \quad \text{if } N > n_1(\epsilon), \qquad (23)$$

for every z in S, for the N in Eq. (23) is necessarily greater than every single least $n_z(\epsilon)$ in Eq. (22).

When an $n_1(\epsilon)$ exists for every positive ϵ, there is a uniform restriction on the least numbers $n_z(\epsilon_k)$ which measure the slowness of convergence of the series throughout S. Hence we say that the series *converges uniformly* in S. The formal definition is:

Definition. The series $\Sigma\, g_n(z)$ *converges uniformly for z in the point set S if, for every positive ϵ, there exists a single integer $n_1(\epsilon)$, depending only on ϵ and not on z, for which Eq. (23) holds.*

The set S is frequently an open or closed region. But it may consist of the points of a curved arc or any infinite set of points. Since a *finite* number of values $n_z(\epsilon)$ is necessarily bounded, no question of uniform convergence can arise for finite sets.

A useful sufficient condition for uniform convergence, which is known as the Weierstrass M-test, is contained in

Theorem 1. Let the positive numbers M_n be so related to the terms of the infinite series $\Sigma\, g_n(z)$ that

$$|g_n(z)| < M_n \qquad (24)$$

for all n and all z in the set S. Then if the infinite series of constants $\Sigma\, M_n$, converges, the series $\Sigma\, g_n(z)$ converges absolutely and uniformly for z in the point set S.

Since $\Sigma\, M_n$ converges, its remainder after N terms must approach zero as N becomes infinite. Hence for any given positive ϵ, there exists an integer $n_1(\epsilon)$ such that

$$\sum_{n=N+1}^{\infty} M_n < \epsilon \quad \text{if } N > n_1(\epsilon). \qquad (25)$$

But for any positive integer m, Eq. (24) implies that

$$\left|\sum_{n=N+1}^{N+m} g_n(z)\right| \leq \sum_{n=N+1}^{N+m} |g_n(z)| \leq \sum_{n=N+1}^{N+m} M_n. \tag{26}$$

In view of Eq. (25), and the theorem in Section 18, on page 37, we may let m become infinite and so obtain

$$\left|\sum_{n=N+1}^{\infty} g_n(z)\right| \leq \sum_{n=N+1}^{\infty} |g_n(z)| \leq \sum_{n=N+1}^{\infty} M_n < \epsilon, \quad \text{if } N > n_1(\epsilon). \tag{27}$$

This holds for all z in S. Hence by Eq. (23) it establishes the uniform convergence of $\Sigma g_n(z)$ stated in the theorem. The absolute convergence follows from the second inequality of Eq. (27).

One consequence of uniform convergence is given by:

Theorem 2. Let the infinite series $\Sigma g_n(z)$ converge uniformly in the point set S to the sum function $G(z)$. Then if each function $g_n(z)$ is continuous in S, the sum function $G(z)$ is also continuous in S.

By Section 13, we must show that for any z_0 in S and $\epsilon > 0$,

$$|G(z) - G(z_0)| = |\Sigma g_n(z) - \Sigma g_n(z_0)| < \epsilon \tag{28}$$

for all z in S which lie sufficiently close to z_0.

From the assumed uniform convergence, for any given positive ϵ_1, there exists an $n_1(\epsilon_1)$ for which Eq. (23) holds. Thus

$$|R_N(z)| < \epsilon_1 \tag{29}$$

for a particular $N > n_1(\epsilon_1)$ and all z in the set S.

With this N, the partial sum $G_N(z)$ defined by Eq. (19) as the sum of a finite number of continuous functions is continuous for all z in the set S. Hence by the definition of continuity,

$$|G_N(z) - G_N(z_0)| < \epsilon_1 \tag{30}$$

for all z of S which also satisfy $|z - z_0| < \delta(\epsilon_1)$. By Eq. (20),

$$G(z) = G_N(z) + R_N(z), \quad G(z_0) = G_N(z_0) + R_N(z_0). \tag{31}$$

It follows from Eqs. (29) to (31) that

$$\begin{aligned}|G(z) - G(z_0)| &= |G_N(z) - G_N(z_0) + R_N(z) - R_N(z_0)| \\ &\leq |G_N(z) - G_N(z_0)| + |R_N(z)| + |R_N(z_0)| \\ &< \epsilon_1 + \epsilon_1 + \epsilon_1 = 3\epsilon_1.\end{aligned} \tag{32}$$

Hence if $3\epsilon_1 = \epsilon$, or $\epsilon_1 = \epsilon/3$, Eq. (28) will hold for all z in S which also satisfy $|z - z_0| < \delta(\epsilon/3)$, as was to be proved.

For the power series $\Sigma a_n z^n$ we shall establish:

Theorem 8. The power series $\Sigma a_n z^n$ converges absolutely and uniformly in any closed circular region $|z| \leq r$, if $r < R$.

Here R is the radius of convergence of Eq. (14). If $R = \infty$, then $r < \infty$ means for any finite value of r.

If R is finite, let $r_1 = \frac{1}{2}(r + R)$. If $R = \infty$, let $r_1 = r + 1$. In either case, $r_1 < R$ and $\Sigma a_n z^n$ converges for $z = r_1$.

Hence
$$|a_n r_1^n| < M \tag{33}$$

as in Eq. (11). Also $r < r_1$, so that if

$$\frac{r}{r_1} = c, \qquad 0 \leq c < 1. \tag{34}$$

And if $|z| \leq r$,

$$\frac{|z|}{r_1} \leq \frac{r}{r_1} = c, \quad \text{or} \quad \left|\frac{z}{r_1}\right| \leq c \text{ where } 0 \leq c < 1. \tag{35}$$

Then for any z in the closed circular region $|z| \leq r$, we have

$$|a_n z^n| = |a_n r_1^n| \left|\frac{z}{r_1}\right|^n \leq M c^n. \tag{36}$$

To apply the Weierstrass M-test, we consider Eq. (36) as Eq. (24) with $g_n(z) = a_n z^n$ and $M_n = M c^n$. Since $0 \leq c < 1$, the geometric series $\Sigma M c^n$ converges. Thus ΣM_n converges, and the M-test applies for $|z| \leq r$. Hence it follows from Theorem 1 that $\Sigma a_n z^n$ converges absolutely and uniformly for $|z| \leq r$, as was to be proved.

Corollary. The sum function of a power series $f(z) = \Sigma a_n z^n$ is a continuous function at any point inside the circle of convergence.

Any point with $|z| < R$ has $|z| < r < R$ where $r = \frac{1}{2}(|z| + R)$ if R is finite, and $r = |z| + 1$ if $R = \infty$. Thus by Theorem 3, the point z is an *interior* point of a region of uniform convergence. But the individual terms $a_n z^n$ are continuous everywhere. Hence by Theorem 2, the continuity of $f(z)$ follows.

21. Differentiation of power series

We next prove:

Theorem. For each value of z inside the circle of convergence of a power series, the sum function of the series $f(z) = \Sigma\, a_n z^n$ has a derivative, which is the sum function of the power series obtained by termwise differentiation:

$$f'(z) = \sum_{n=1}^{\infty} n a_n z^{n-1}. \tag{37}$$

Let z be inside the circle of convergence. Then if $|z| = r$, $r < R$. For R finite, let $p = \frac{1}{4}(R - r)$, $r_1 = r + 2p$. For R infinite, let $p = 1$, $r_1 = r + 2$. In either case, $r + p < r_1 < R$, and by Theorem 3 of Section 20, the series $\Sigma\, a_n z^n$ converges absolutely and uniformly for $|z| \leq r_1$. Hence in particular $\Sigma |a_n| r^n$ and $\Sigma |a_n|(r + p)^n$ each converge, as well as the series obtained by termwise subtraction. Its general term is pM_n, where

$$M_n = |a_n| \frac{(r+p)^n - r^n}{p} = |a_n|\left(nr^{n-1} + \frac{n(n-1)}{2!} r^{n-2} p + \ldots\right). \tag{38}$$

The last form shows that it is positive and decreases if p is replaced by any smaller number.

From now on we keep z fixed as the value with $|z| = r$ with which we started. We let $\Delta z = h$ be any complex number with $|h| < p$. And we form the series for the difference quotient

$$G(h) = \frac{f(z + h) - f(z)}{h}$$

$$= \sum_{n=0}^{\infty} a_n \frac{(z+h)^n - z^n}{h} = \sum_{n=0}^{\infty} g_n(h). \tag{39}$$

Its general term is

$$g_n(h) = a_n \frac{(z+h)^n - z^n}{h}$$

$$= a_n\left(nz^{n-1} + \frac{n(n-1)}{2!} z^{n-2} h + \ldots\right). \tag{40}$$

The absolute value of this is less than the term on the right of Eq. (38), which was the term of a convergent series of positive constants independent of h. Hence the Weierstrass M-test applies, with h as the variable in the role previously played by z, on the set $|h| < p$. Theorem 1 shows that $\Sigma\, g_n(h)$ converges uniformly in the complex variable h for $|h| < p$. The last form in Eq. (40) shows that the $g_n(h)$

are polynomials in h and so are continuous functions of h, with $g_n(0) = a_n n z^{n-1}$. Hence by Theorem 2, the sum function is continuous for $|h| < p$, and in particular for $h = 0$. Hence

$$\lim_{h \to 0} G(h) = G(0) = \sum_{n=0}^{\infty} g_n(0) = \sum_{n=1}^{\infty} a_n n z^{n-1}. \qquad (41)$$

It follows from this and Eq. (39) that

$$f'(z) = \lim_{h \to 0} \frac{f(z+h) - f(z)}{h} = \sum_{n=1}^{\infty} a_n n z^{n-1}. \qquad (42)$$

Thus the derivative of $f(z) = \Sigma a_n z^n$ exists, and has the value stated in the theorem.

Corollary 1. The sum function of a power series with $R > 0$ is analytic for all values of z with $|z| < R$.

This follows from the definition of an analytic function given in Section 14. If $R = \infty$, the sum function is analytic everywhere and we say that the function is an *entire* function.

Corollary 2. The radius of convergence of the derived series $\Sigma a_n n z^{n-1}$ is the same as that of the original series $\Sigma a_n z^n$.

Let R_D be the radius of convergence of the derived series. Then from Eq. (14), we have $R_D = 1/A_D$, where

$$A_D = \overline{\lim_{n \to \infty}} \sqrt[n]{n|a_n|} = \lim_{n \to \infty} \sqrt[n]{n} \, \overline{\lim_{n \to \infty}} \sqrt[n]{|a_n|}. \qquad (43)$$

But as n becomes infinite, $\log \sqrt[n]{n} = (\log n)/n$ approaches zero, so that $\sqrt[n]{n} = e^{(\log n)/n}$ approaches $e^0 = 1$. Hence from Eq. (43),

$$A_D = A, \quad \text{and} \quad R_D = \frac{1}{A_D} = \frac{1}{A} = R. \qquad (44)$$

This proves Corollary 2.

Corollary 3. For the z of the theorem with $|z| < R$, the function $f(z) = \Sigma a_n z^n$ has derivatives of all orders. The mth derivative,

$$f^{(m)}(z) = \sum_{n=m}^{\infty} n(n-1) \ldots (n-m+1) a_n z^{n-m}, \qquad (45)$$

is given by termwise differentiation m times. And for every m the derived series of Eq. (45) has the same radius of convergence as the original series.

That the radius of convergence of the series in Eq. (45) is R follows by an m-fold application of Corollary 2. Hence any z with $|z| < R$ is inside the circle of convergence for each derived series, so that the theorem may be applied m times to prove that Eq. (45) holds.

Corollary 4. If $f(z) = \Sigma\, a_n z^n$ has $R > 0$, the expansion is necessarily a Maclaurin series of $f(z)$ with

$$a_n = \frac{f^{(n)}(0)}{n!}. \tag{46}$$

Since $0 < R$, we may put $z = 0$ in Eq. (45). For $z = 0$, the power series reduces to its first or constant term with $n = m$. Division by $m!$ leads to Eq. (46) with m in place of n.

Most of the properties which we have discussed for the power series $\Sigma\, a_n z^n$ hold with only minor modification for the more general power series

$$F(z) = \Sigma\, a_n (z - b)^n. \tag{47}$$

Thus this series converges for $|z - b| < R$, where R is given by Eq. (14). For any z inside this circle of convergence, the sum function $F(z)$ is analytic. Its derivative of the mth order is

$$F^{(m)}(z) = \sum_{n=m}^{\infty} n(n-1)\ldots(n-m+1)a_n(z-b)^{n-m}. \tag{48}$$

This series has the same radius of convergence R as that of Eq. (47), and if $R > 0$, the series of Eq. (47) is a Taylor's series with

$$a_n = \frac{F^{(n)}(b)}{n!}. \tag{49}$$

These properties all follow from the fact that if $Z = z - b$, $z = Z + b$, and Eq. (47) is equivalent to

$$F(z) = F(Z + b) = \Sigma\, a_n Z^n. \tag{50}$$

Exercise 10

1. Prove that the geometric series $\Sigma\, z^n$ is not uniformly convergent in the domain of convergence $|z| < 1$, by studying the behavior of $R_N = z^{N+1}/(1 - z)$ for z near 1. For example, note that for any *fixed* N, $\lim_{z \to 1} R_N = \infty$.

2. Prove that the series $\sum z^n/n^2$ is uniformly convergent in its region of convergence $|z| \leq 1$.

For each given value of $g_n(z)$, verify the expression for the partial sum S_N of the series $\sum_{n=1}^{\infty} g_n(z)$. Show that the series converges for all values of z to the expression given for $G(z)$. Also prove that in any neighborhood of $z = 0$ the convergence is not uniform, either by noticing the discontinuity of $G(z)$, or by directly studying the behavior of the remainder R_N for fixed N as $z \to 0$.

3. $g_n(z) = \dfrac{1}{1 + n|z|} - \dfrac{1}{1 + (n+1)|z|}$,

$S_N = \dfrac{1}{1 + |z|} - \dfrac{1}{1 + (N+1)|z|}$,

$G(z) = \dfrac{1}{1 + |z|}$ if $z \neq 0$, $G(0) = 0$.

4. $g_n(z) = \dfrac{1}{(1 + |z|)^{n-1}} - \dfrac{1}{(1 + |z|)^n}$,

$S_N = 1 - \dfrac{1}{(1 + |z|)^N}$,

$G(z) = 1$ if $z \neq 0$, $G(0) = 0$.

5. Assume it known that the Cauchy product rule

$$\sum_{n=0}^{\infty} (c_0 C_n + c_1 C_{n-1} + \ldots + c_n C_0) = \left(\sum_{n=0}^{\infty} c_n\right)\left(\sum_{n=0}^{\infty} C_n\right)$$

is valid for absolutely convergent series with real terms. Prove that the rule is valid for absolutely convergent series of complex terms. *Hint:* Expand $(a + ib)(A + iB) = aA - bB + i(aB + bA)$.

6. Use the result of Problem 5 to show that if $f(z) = \sum a_n z^n$ for $|z| < R_1$ and $g(z) = \sum b_n z^n$ for $|z| < R_2$, then for any z with $|z| = r$ where $r < R_1$ and $r < R_2$,

$f(z) g(z) = \sum c_n z^n$, where $c_n = (a_0 b_n + a_1 b_{n-1} + \ldots + a_n b_0)$.

7. Verify directly that Problem 6 gives a correct result if $f(z) = 1 - z$, $g(z) = 1/(1 - z) = 1 + z + z^2 + \ldots + z^n + \ldots$, $f(z)g(z) = 1$. This shows that the product *may* have a larger R than either series.

8. In Problem 6, let
$$A_1 = \overline{\lim_{n \to \infty}} \sqrt[n]{|a_n|} \quad \text{and} \quad A_2 = \overline{\lim_{n \to \infty}} \sqrt[n]{|b_n|}.$$
Show that if
$$A_3 = \overline{\lim_{n \to \infty}} \sqrt[n]{|c_n|}, \quad A_3 \leq \overline{\lim_{n \to \infty}} \sqrt[n]{(n+1)} A_1,$$
if $A_1 \geq A_2$. Deduce directly from this that for the radii of convergence of the three series, $R_3 \geq R_1$ and $R_3 \geq R_2$. This fact is implied by Problem 6. The *inequality* occurs in Problem 7.

9. By consideration of $R_N = \dfrac{1}{z - b}\left[1 - \left(\dfrac{z}{b}\right)^{N+1}\right]$, show that
$$\frac{1}{z - b} = \left(-\frac{1}{b}\right)\left(1 + \frac{z}{b} + \frac{z^2}{b^2} + \ldots + \frac{z^n}{b^n} + \ldots\right),$$
for $|z| < |b|$.

10. Deduce the binomial expansion for negative integral powers by differentiating the expansion of Problem 9, $(m - 1)$ times.
$$\frac{1}{(z-b)^m} = \left[-\frac{1}{b}\right]^m \left[1 + \frac{mz}{b} + \ldots \right.$$
$$\left. + \frac{(n+m-1)(n+m-2)\ldots(n+1)}{(m-1)!} \cdot \frac{z^n}{b^n} + \ldots\right].$$

11. Verify that the expansion of Problem 10 is a Maclaurin expansion by calculating directly $f^{(n)}(0)/n!$.

12. Let the two power series $\Sigma a_n z^n$ and $\Sigma b_n z^n$ each converge for $|z| < r_1$ where $r_1 > 0$. Show that if they have the same sum for any infinite sequence of points $z_n \neq 0$ but such that $z_n \to 0$ as $n \to \infty$, then $a_n = b_n$. *Hint:* For all z_n,
$$a_0 + a_1 z + a_2 z^2 + \ldots = b_0 + b_1 z + b_2 z^2 + \ldots.$$
Letting $z_n \to 0$ gives $a_0 = b_0$. Hence for the $z_n \neq 0$,
$$a_1 + a_2 z + \ldots = b_1 + b_2 z + \ldots;$$
Letting $z_n \to 0$ gives $a_1 = b_1$, and repetition of the process gives $a_n = b_n$ for all n.

Given that $f(z) = \Sigma a_n z^n$ has $R > 0$. Use Problem 12 to show that

13. If $f(1/n) = 2/n$, then $f(z) = 2z$.
14. If $f(1/n) = 0$, then $f(z) = 0$.

15. If $f\left(\dfrac{1}{n}\right) = \dfrac{n}{n+1}$, then $f(z) = \dfrac{1}{1+z} = \Sigma\,(-1)^n z^n$.

16. It is impossible to have
$$f\left(\frac{1}{2n+1}\right) = \frac{1}{2n+1}, \quad f\left(\frac{1}{2n}\right) = \frac{1}{n}.$$

17. Let $g(x) = \Sigma\, b_n x^n$ for all real x such that $0 < x < p$. Show that the only $f(z) = \Sigma\, a_n z^n$ with $R > 0$ such that $f(x) = g(x)$ for $0 < x < p$ has $a_n = b_n$ and $R \geq p$. When the b_n are real, $f(z)$ is called the *extension* or *continuation* of $f(x)$ into the complex plane.

4

THE ELEMENTARY FUNCTIONS

In this chapter we define the exponential, trigonometric, and hyperbolic functions for complex values of the variable. We develop their basic properties, and either as our definition or in the course of the discussion we obtain power series expansions for these functions which converge everywhere. We also study the inverses of these functions. In connection with these multiple-valued functions, we present a brief introduction to the study of Riemann surfaces.

22. The exponential function

The expansion of e^x

$$e^x = 1 + x + \frac{x^2}{2!} + \ldots + \frac{x^n}{n!} + \ldots, \tag{1}$$

in a Maclaurin series is valid for all real values of x. This leads us to consider the series

$$1 + z + \frac{z^2}{2!} + \ldots + \frac{z^n}{n!} + \ldots \tag{2}$$

for complex values of z. Here

$$a_n = \frac{1}{n!}, \quad \frac{a_{n+1}}{a_n} = \frac{n!}{(n+1)!} = \frac{1}{n+1}, \quad \lim_{n \to \infty} \left|\frac{a_{n+1}}{a_n}\right| = 0. \tag{3}$$

Hence in Eq. (17) of Section 19, $L = 0$ and $R = \infty$. The series (2) therefore defines a sum function which is analytic for all complex values of z, or is an *entire* function. As we shall soon verify, for all complex values this entire function satisfies the same basic laws as the real exponential function e^x. This fact motivates the notation e^z or $\exp z$ for the sum function of the series (2).

Thus the exponential function e^z is *defined* for complex exponents in a single-valued manner by the expansion

$$e^z = \sum_{n=0}^{\infty} \frac{z^n}{n!}, \qquad (4)$$

We recall that $0! = 1$. For a real value of the complex variable z, $z = x$, Eq. (4) reduces to Eq. (1). Thus the definition of e^z is consistent with that of e^x used in algebra and calculus. By Problem 17 of Exercise 10, no other power series in z could have a sum function equal to e^x for all real values. In this sense, the power series (4) continues the function e^x from the real axis into the complex plane.

From the absolute convergence of the series involved, we may prove as in Problem 1, of Exercise 11 that

$$\left(\sum_{n=0}^{\infty} \frac{z_1^n}{n!} \right) \left(\sum_{n=0}^{\infty} \frac{z_2^n}{n!} \right) = \sum_{n=0}^{\infty} \frac{(z_1 + z_2)^n}{n!}. \qquad (5)$$

This shows that the *addition theorem*

$$e^{z_1} e^{z_2} = e^{z_1 + z_2} \qquad (6)$$

holds for complex as well as for real z. Consequently

$$\frac{e^{z_1}}{e^{z_2}} = e^{z_1 - z_2}, \quad \text{since } e^{z_2} e^{z_1 - z_2} = e^{z_1}. \qquad (7)$$

Next we observe that each term of the series (4) has as its derivative the preceding term. Thus the series reproduces itself when differentiated termwise, and by the theorem of Section 21,

$$\frac{de^z}{dz} = e^z. \qquad (8)$$

From this and Section 15, we may deduce further that

$$\frac{de^w}{dz} = e^w \frac{dw}{dz} \quad \text{and} \quad \frac{de^{Kz}}{dz} = Ke^{Kz}. \qquad (9)$$

It follows that the rules for differentiation of complex exponentials are similar to those found in the calculus of reals.

To find an expression for e^z in closed form, we proceed as follows. Let $z = x + iy$. Then it follows from Eq. (6) that

$$e^z = e^{x+iy} = e^x e^{iy}. \tag{10}$$

But from Eq. (4) we have for real values of y,

$$e^{iy} = \sum_{n=0}^{\infty} \frac{(iy)^n}{n!} = \sum_{m=0}^{\infty} (-1)^m \frac{y^{2m}}{(2m)!} + i \sum_{m=0}^{\infty} (-1)^m \frac{y^{2m+1}}{(2m+1)!}. \tag{11}$$

But as recalled below in Eq. (26), the two *real* power series in Eq. (11) are the Maclaurin series expansions of $\cos y$ and $\sin y$, respectively. This proves the *Euler relation*:

$$e^{iy} = \cos y + i \sin y = \operatorname{cis} y. \tag{12}$$

From Eqs. (10) and (12) we obtain

$$e^z = e^{x+iy} = e^x(\cos y + i \sin y). \tag{13}$$

This expression enables us to calculate the value of e^z for a given z, since for a given numerical value of x and y, the real functions e^x, $\cos y$, and $\sin y$ (for y in radians) can be evaluated by using suitable tables. As indicated in Problem 2 of Exercise 11, Eq. (13) implies Eq. (4). Hence it is a possible alternative definition of e^z.

As particular values obtained from Eq. (12), we note that

$$e^{2\pi i} = 1, \quad e^{\pi i} = -1, \quad e^{\pi i/2} = i, \quad e^{-\pi i/2} = -i. \tag{14}$$

And for any *real* value of y, Eq. (12) shows that

$$|e^{iy}| = |\operatorname{cis} y| = \sqrt{\cos^2 y + \sin^2 y} = 1. \tag{15}$$

From Eqs. (10) and (15) it follows that

$$|e^z| = |e^x||e^{iy}| = e^x \quad \text{or} \quad |e^z| = e^{\operatorname{Re} z}, \tag{16}$$

since e^x is positive for any real x.

Also from Section 7 and Eq. (13) we see that

$$\arg e^z = y = \operatorname{Im} z. \tag{17}$$

From Eq. (6) and the first relation in Eq. (14) we have

$$e^{z+2\pi i} = e^z e^{2\pi i} = e^z. \tag{18}$$

Thus the function e^z is *periodic* with the period $2\pi i$. Also

$$e^{z-2\pi i} = e^z, \quad e^{z+2k\pi i} = e^z, \tag{19}$$

where k is any positive or negative integer, or zero. Thus e^z has the same value at all points of the z-plane which result from one point by a single or repeated application of the translation $2\pi i$, or $-2\pi i$.

Conversely, we may show that if

$$e^{z_1} = e^{z_2}, \quad z_2 = z_1 + 2k\pi i, \tag{20}$$

for, from Eq. (7) and the first relation of Eq. (20), we have

$$e^{z_2 - z_1} = \frac{e^{z_2}}{e^{z_1}} = \frac{e^{z_2}}{e^{z_2}} = 1. \tag{21}$$

From Eqs. (16) and (13), the relation $e^z = e^{x+iy} = 1$ implies that $e^x = 1$, $\cos y = 1$, $\sin y = 0$. This leads to $x = 0$, $y = 2k\pi$, so that $z = 2k\pi i$. Hence Eq. (21) leads to $z_2 - z_1 = 2k\pi i$, or the second relation of Eq. (20).

Suppose that w is a value assumed anywhere by the function $w = e^z$. Then the statement following Eq. (19) shows that w is assumed by e^z for some value of z in the infinite strip

$$-\pi < y \leq \pi \quad \text{or} \quad -\pi < \operatorname{Im} z \leq \pi. \tag{22}$$

Any strip obtained from this by a parallel translation would have the same property. We call any such strip, and in particular that of Eq. (22), a *fundamental region* for the function e^z.

The function e^z *never* assumes the value zero, for it follows from Eq. (6) that for any z,

$$e^z e^{-z} = e^0 = 1. \tag{23}$$

Hence $e^z \neq 0$, since a zero factor would make the product zero.

In the *fundamental strip* (22) the function e^z assumes each *nonzero* value w at *one and only one point*. Let $w = u + iv$. Then if r_1, θ_1 are polar coordinates for the point u,v as in Section 7 we have $w = r_1 \operatorname{cis} \theta_1$. Since $w \neq 0$, $r_1 \neq 0$. Hence if

$$z = \log r_1 + i(\theta_1 + 2k\pi), \tag{24}$$

we have

$$e^z = e^{\log r_1} e^{i\theta_1} e^{2k\pi i} = r_1 \operatorname{cis} \theta_1 = w. \tag{25}$$

One of the values of Eq. (24) will lie in the fundamental strip. The value w could not be taken on for two distinct values of z because of Eq. (20).

Exercise 11

1. Show that the Cauchy product rule proved in Problem 5 of Exercise 10 applies to the product in Eq. (5), and that the general term of the product is

$$\sum_{k=0}^{n} \frac{z_1^{n-k}}{(n-k)!} \frac{z_2^k}{k!} = \frac{1}{n!} \sum_{k=0}^{n} \frac{n!}{(n-k)!k!} z_1^{n-k} z_2^k = \frac{(z_1+z_2)^n}{n!}$$

by the binomial theorem. This proves Eq. (5).

2. From the Maclaurin series for e^x, $\cos y$, and $\sin y$, and the second equality of Eq. (11) read backward, deduce that

$$e^x(\cos y + i \sin y) = \sum_{n=0}^{\infty} \frac{x^n}{n!} \sum_{n=0}^{\infty} \frac{(iy)^n}{n!}$$

Finally deduce from Eq. (5) that this last product is $\sum_{n=0}^{\infty} z^n/n!$.

3. Use Eq. (9) and the rule for differentiating a product to verify that

$$\frac{d}{dz}(e^z e^{K-z}) = e^z e^{K-z} + e^z(-e^{K-z}) = 0.$$

Deduce from Theorem 2 of Section 16 that $e^z e^{K-z} = C$. For $z = 0$, $C = e^K$. With $z = z_1$, $K = z_1 + z_2$, this gives an alternative derivation of Eq. (6).

Check each of the following evaluations of $\exp z = e^z$.

4. $\exp \dfrac{3-\pi i}{2} = -e^{3/2} i$.   5. $\exp(4+3\pi i) = -e^4$.

6. $\exp \dfrac{2+\pi i}{4} = \sqrt{\dfrac{e}{2}}(1+i)$.   7. $\exp \dfrac{3-\pi i}{3} = \dfrac{e}{2}(1-i\sqrt{3})$.

Verify that

8. If $\exp z = -10$, $z = \log 10 + (1+2k)\pi i$.
9. If $\exp z = 2 - 2i\sqrt{3}$, $z = \log 4 + (-\tfrac{1}{3} + 2k)\pi i$.
10. $\exp z$ is real if and only if $y = \operatorname{Im} z = k\pi$.
11. $\exp z$ is pure imaginary if and only if
$$y = \operatorname{Im} z = (\tfrac{1}{2} + k)\pi.$$
12. $|\exp(-3iz + 5i)| = e^{3y}$.   13. $|\exp(4i - z)| = e^{-x}$.

14. Show that $\exp \bar{z} = \overline{\exp z}$. From $\dfrac{\partial}{\partial \bar{z}}(\exp \bar{z}) = \exp \bar{z} \neq 0$, and

Problem 24 of Exercise 7, deduce that this function is nowhere analytic.

15. Use the considerations of Section 16, in particular Theorem 4 and Eq. (38) to show directly that the function $f(z) = e^x(\cos y + i \sin y)$ has a derivative equal to itself for every value of $z = x + iy$.

Use the relation of Problem 23 of Exercise 8, $f(z) = \int [u_x(z,0) - iu_y(z,0)]\, dz$, to show that

16. If $u = e^x \cos y$, $f(z) = \int e^z\, dz = e^z + Ci$.
17. If $u = e^{-x} \sin y$, $f(z) = \int -ie^{-z}\, dz = ie^{-z} + Ci$.
18. If $u = e^x(x \cos y - y \sin y)$, $f(z) = \int e^z(z + 1)\, dz = ze^z + Ci$.

23. The trigonometric functions

The Maclaurin series

$$\cos x = 1 - \frac{x^2}{2!} + \frac{x^4}{4!} - \ldots + (-1)^m \frac{x^{2m}}{(2m)!} + \ldots, \tag{26}$$

$$\sin x = x - \frac{x^3}{3!} + \frac{x^5}{5!} - \ldots + (-1)^m \frac{x^{2m+1}}{(2m+1)!} + \ldots,$$

are valid for all real values of x. We use them to continue the functions $\cos x$ and $\sin x$ from the real axis into the complex plane by *defining*

$$\cos z = \sum_{m=0}^{\infty} (-1)^m \frac{z^{2m}}{(2m)!}, \quad \sin z = \sum_{m=0}^{\infty} (-1)^m \frac{z^{2m+1}}{(2m+1)!}. \tag{27}$$

For either series, the ratio of the term in z^{n+2} to the preceding term in z^n is in absolute value

$$\left| \frac{z^{n+2}}{(n+2)!} \frac{n!}{z^n} \right| = \frac{|z^2|}{(n+1)(n+2)}. \tag{28}$$

For any fixed z, this approaches zero when n becomes infinite through either even numbers $2m$ or odd numbers $(2m + 1)$. As the limit zero is less than unity, it follows from the ratio test for real series that both series of Eq. (27) converge absolutely for all z, and hence have $R = \infty$. Thus each of the functions $\cos z$ and $\sin z$ is an *entire* function, analytic for all complex values of z.

We see from the series that $\cos z$ is an *even* function,
$$\cos(-z) = \cos z, \tag{29}$$
while $\sin z$ is an *odd* function,
$$\sin(-z) = -\sin z. \tag{30}$$
If we replace y by z in Eq. (11), and use Eq. (27), we obtain
$$e^{iz} = \cos z + i \sin z. \tag{31}$$
Replacing z by $-z$, and using Eqs. (29) and (30), we find
$$e^{-iz} = \cos z - i \sin z. \tag{32}$$
We may solve Eqs. (31) and (32) for $\sin z$ and $\cos z$ in the form
$$\cos z = \frac{e^{iz} + e^{-iz}}{2}, \quad \sin z = \frac{e^{iz} - e^{-iz}}{2i}. \tag{33}$$
The relations of Eqs. (31) to (33) are known as *Euler's formulas*. It follows from Eqs. (33) and (6) that
$$\cos(z_1 + z_2) = \frac{e^{iz_1}e^{iz_2} + e^{-iz_1}e^{-iz_2}}{2}, \tag{34}$$
$$\sin(z_1 + z_2) = \frac{e^{iz_1}e^{iz_2} - e^{-iz_1}e^{-iz_2}}{2i}.$$
Then Eqs. (31) and (32) show that
$$e^{iz_1}e^{iz_2} = (\cos z_1 + i \sin z_1)(\cos z_2 + i \sin z_2), \tag{35}$$
$$e^{-iz_1}e^{-iz_2} = (\cos z_1 - i \sin z_1)(\cos z_2 - i \sin z_2).$$
On reducing the right members of Eq. (34) by means of Eq. (35), and simplifying, we obtain the *addition theorems*:
$$\cos(z_1 + z_2) = \cos z_1 \cos z_2 - \sin z_1 \sin z_2, \tag{36}$$
$$\sin(z_1 + z_2) = \sin z_1 \cos z_2 + \cos z_1 \sin z_2.$$
Let $z = x + iy$. Then from Eq. (36) we have
$$\cos(x + iy) = \cos x \cos iy - \sin x \sin iy, \tag{37}$$
$$\sin(x + iy) = \sin x \cos iy + \cos x \sin iy.$$
But from Eq. (33), we have
$$\cos iy = \frac{e^y + e^{-y}}{2}, \quad \sin iy = \frac{e^{-y} - e^y}{2i} = i\frac{e^y - e^{-y}}{2}. \tag{38}$$

We recall that for a real value y, the hyperbolic cosine and hyperbolic sine are defined by the relations

$$\cosh y = \frac{e^y + e^{-y}}{2}, \quad \sinh y = \frac{e^y - e^{-y}}{2} \tag{39}$$

Then we may deduce from Eqs. (38) and (39) that

$$\cos iy = \cosh y, \quad \sin iy = i \sinh y. \tag{40}$$

We may use this to reduce Eq. (37) to the form:

$$\cos(x + iy) = \cos x \cosh y - i \sin x \sinh y,$$
$$\sin(x + iy) = \sin x \cosh y + i \cos x \sinh y. \tag{41}$$

These expressions enable us to calculate the values of $\sin z$ and $\cos z$ for a given z, since the real functions $\cos x$, $\sin x$, $\cosh y$, $\sinh y$ can be evaluated by using suitable tables.

For a real value of the complex variable z, $z = x$, Eq. (27) reduces to Eq. (26). Thus the values of $\cos z$ and $\sin z$ must be $\cos x$ and $\sin x$ when $y = 0$. But from Eq. (39), $\cosh 0 = 1$ and $\sinh 0 = 0$, so that when $y = 0$, the right members of Eq. (41) do reduce to $\cos x$ and $\sin x$. In particular we note the values

$$\cos 0 = 1, \quad \cos \frac{\pi}{2} = 0, \quad \cos \pi = -1, \quad \cos 2\pi = 1,$$
$$\sin 0 = 0, \quad \sin \frac{\pi}{2} = 1, \quad \sin \pi = 0, \quad \sin 2\pi = 0. \tag{42}$$

From Eqs. (36), (29), and (30) we may deduce that

$$\cos(z_1 - z_2) = \cos z_1 \cos z_2 + \sin z_1 \sin z_2,$$
$$\sin(z_1 - z_2) = \sin z_1 \cos z_2 - \cos z_1 \sin z_2. \tag{43}$$

If $z_1 = z_2 = z$ in the first equation, we have

$$\cos^2 z + \sin^2 z = 1. \tag{44}$$

The special values of Eq. (42), and Eqs. (36) and (43) imply

$$\sin\left(\frac{\pi}{2} - z\right) = \cos z, \quad \cos(z + \pi) = -\cos z, \tag{45}$$

and a number of similar relations. In particular, we have

$$\cos(z + 2\pi) = \cos z, \quad \sin(z + 2\pi) = \sin z, \tag{46}$$

so that $\cos z$ and $\sin z$ are each *periodic* with the period 2π.

58 THE ELEMENTARY FUNCTIONS Chap. 4

Our Eqs. (29), (30), (36), and (42) to (46) show that the fundamental equations of real trigonometry are also valid for complex variables. As the elementary identities of real trigonometry are algebraic deductions from these, all such identities must continue to hold for the trigonometric functions of a complex variable.

It follows from Eq. (39) that

$$\cosh^2 y = 1 + \sinh^2 y. \tag{47}$$

From this and Eq. (44), we may deduce from Eq. (41) that

$$\begin{aligned}
|\cos(x+iy)|^2 &= \cos^2 x \cosh^2 y + \sin^2 x \sinh^2 y \\
&= \cos^2 x (1 + \sinh^2 y) + (1 - \cos^2 x) \sinh^2 y \\
&= \cos^2 x + \sinh^2 y, \\
|\sin(x+iy)|^2 &= \sin^2 x \cosh^2 y + \cos^2 x \sinh^2 y \\
&= \sin^2 x (1 + \sinh^2 y) + (1 - \sin^2 x) \sinh^2 y \\
&= \sin^2 x + \sinh^2 y.
\end{aligned} \tag{48}$$

These equations show that

$$|\cos z|^2 = \cos^2 x + \sinh^2 y, \quad |\sin z|^2 = \sin^2 x + \sinh^2 y. \tag{49}$$

Since $|\sinh y|$ increases from 0 to infinity as $|y|$ increases from 0 to infinity, it follows that both $|\cos z|$ and $|\sin z|$ can be made arbitrarily large by taking z sufficiently far from the real axis. Thus the inequalities $|\cos x| \leq 1$, $|\sin x| \leq 1$ which hold for reals do *not* hold for unrestricted values of the complex arguments.

A value z_1 is called a *zero* of $f(z)$ if $f(z_1) = 0$. If $\cos z = 0$, $|\cos z| = 0$ and it follows from Eq. (49) that $\cos x = 0$, $\sinh y = 0$. Hence $y = 0$ and $x = \pi/2 + k\pi$. Again, if $\sin z = 0$, $|\sin z| = 0$ and it follows from Eq. (49) that $\sin x = 0$, $\sinh y = 0$. Hence $y = 0$ and $x = k\pi$. Thus the only *zeros* of $\cos z$ and $\sin z$ are the *real zeros* of $\cos x$ and $\sin x$ implied by

$$\cos(\pi/2 + k\pi) = 0, \quad \sin k\pi = 0. \tag{50}$$

By the theorem of Section 21, we may apply termwise differentiation to each of the series (27) and so deduce that

$$\frac{d(\cos z)}{dz} = -\sin z, \quad \frac{d(\sin z)}{dz} = \cos z. \tag{51}$$

These relations could also be derived from Eqs. (33) and (9).

Sec. 23   THE ELEMENTARY FUNCTIONS   59

The other trigonometric functions are defined by

$$\tan z = \frac{\sin z}{\cos z}, \quad \cot z = \frac{\cos z}{\sin z}, \quad \sec z = \frac{1}{\cos z}, \quad \csc z = \frac{1}{\sin z}, \quad (52)$$

By Section 15, each of these is analytic except for the real zeros of the denominator implied by Eq. (50). Their derivatives are found in Problems 16 and 17 of Exercise 12.

Exercise 12

1. By multiplication of Eqs. (31) and (32), deduce that $e^{iz}e^{-iz} = (\cos z + i \sin z)(\cos z - i \sin z)$, and use this to give an alternative derivation of Eq. (44).

Use Eqs. (31) to (33) to give a direct proof of each identity.

2. $\cos z_2 - \cos z_1 = 2 \sin \dfrac{z_1 + z_2}{2} \sin \dfrac{z_1 - z_2}{2}$.

3. $\sin z_2 - \sin z_1 = 2 \cos \dfrac{z_1 + z_2}{2} \sin \dfrac{z_2 - z_1}{2}$.

4. Show that Eqs. (33) and (4) imply Eq. (27), so that we could take Eq. (33) as the definition of $\cos z$ and $\sin z$.

5. That $e^{ix} = \operatorname{cis} x$ is known for real x from Eq. (12). From this and Eqs. (41) and (39), deduce that if $z = x + iy$,

$$\cos z = \tfrac{1}{2}[e^y \operatorname{cis}(-x) + e^{-y} \operatorname{cis} x] = \tfrac{1}{2}(e^{iz} + e^{-iz}),$$

$$\sin z = \frac{i}{2}[e^y \operatorname{cis}(-x) - e^{-y} \operatorname{cis} x] = \frac{1}{2i}(e^{iz} - e^{-iz}).$$

Since this implies Eq. (33), by Problem 4, we could take Eq. (41) as the definition of $\cos z$ and $\sin z$.

6. Show that $|\cos z|^2 = \cosh^2 y - \sin^2 x$. From this and Eq. (49) deduce that

$$\sinh |y| \leq |\cos z| \leq \cosh y.$$

7. Show that $|\sin z|^2 = \cosh^2 y - \cos^2 x$. From this and Eq. (49) deduce that

$$\sinh |y| \leq |\sin z| \leq \cosh y.$$

8. From Problem 2 and Eq. (50) deduce that $\cos z_2 = \cos z_1$ if and only if $z_2 = \pm z_1 + 2k\pi$.

9. From Problem 3 and Eq. (50) deduce that $\sin z_2 = \sin z_1$ if and only if $z_2 = z_1 + 2k\pi$ or $z_2 = -z_1 + (2k+1)\pi$.

10. If $W = \cos z$, deduce that $(e^{iz})^2 - 2We^{iz} + 1 = 0$ and hence $e^{iz} = W \pm \sqrt{W^2 - 1} = w_1, w_2$. Here $w_1 \neq w_2$ unless $W = \pm 1$, and neither value is zero since $w_1 w_2 = 1$.

11. If $W = \sin z$, deduce that $(e^{iz})^2 - 2iWe^{iz} - 1 = 0$ and hence $e^{iz} = iW \pm \sqrt{1 - W^2} = w_1, w_2$. Here $w_1 \neq w_2$ unless $W = \pm 1$. And neither value is zero since $w_1 w_2 = -1$.

12. Show that $-\pi < x \leq \pi$ is a *fundamental region* for each of the functions $\cos z$ and $\sin z$, in which the functions take on 1 and -1 at one point, and any value $W \neq \pm 1$ at precisely two points. *Hint:* $\cos 0 = 1$, $\cos \pi = -1$, $\sin \pi/2 = 1$, $\sin (-\pi/2) = -1$. For $W \neq \pm 1$, by Problems 10 and 11, there are two distinct nonzero values w_1 and w_2 of e^{iz}. Since Im $(iz) = x$, by the discussion following Eq. (22), there are two distinct values z_1 and z_2 for which $e^{iz_1} = w_1$, $e^{iz_2} = w_2$ in the strip $-\pi < x \leq \pi$. There are no additional points in the strip, by Problems 8 and 9.

13. Find all the values of z for which $\cos z = 2$. *Hint:* From Problem 10, $e^{iz} = 2 \pm \sqrt{3} = (2 \pm \sqrt{3})$ cis 0 so that $x = 2k\pi$, $y = -\log (2 \pm \sqrt{3})$. For $x = 2k\pi$, from Eq. (41), $2 = \cosh y$, $y = \pm \cosh^{-1} 2$. Hence $z = 2k\pi \pm 1.317i$.

14. Find all the values of z for which $\sin z = 2$. *Hint:* From Problem 11, $e^{iz} = i(2 \pm \sqrt{3}) = (2 \pm \sqrt{3})$ cis $(\pi/2)$ so that $x = \pi/2 + 2k\pi$, $y = -\log (2 \pm \sqrt{3})$. For $x = \pi/2 + 2k\pi$, from Eq. (41), $2 = \cosh y$, $y = \pm\cosh^{-1} 2$. Hence $z = \pi/2 + 2k\pi \pm 1.317i$.

15. Let $W = \cos z$ lead to z_1 and z_2 as in Problem 12. From Problem 10, $e^{iz_1} e^{iz_2} = w_1 w_2 = 1$. Deduce that $z_2 = -z_1$. For the same $W = \sin Z$, let the two values of Problem 12 be Z_1 and Z_2. From Problems 10 and 11, $e^{iZ} = ie^{iz}$, deduce that if Re $z_1 \geq 0$, $\pi/2 \pm z_1$ are possible values of Z_1, Z_2. Compare the results of Problems 13 and 14 with $k = 0$.

Use the rule for differentiating quotients together with Eqs. (52), (51), (44) and (50) to verify that

16. $\dfrac{d(\tan z)}{dz} = \sec^2 z$, $\dfrac{d(\sec z)}{dz} = \sec z \tan z$, if $z \neq \pi/2 + k\pi$.

17. $\dfrac{d(\cot z)}{dz} = -\csc^2 z,\ \dfrac{d(\csc z)}{dz} = -\csc z \cot z$, if $z \neq k\pi$.

Use Problem 24 of Exercise 7 to show that:

18. $\cos \bar{z}$ and $\sec \bar{z}$ are differentiable only at $z = k\pi$.
19. $\sin \bar{z}$ and $\csc \bar{z}$ are differentiable only at $z = \pi/2 + k\pi$.
20. $\tan \bar{z}$ and $\cot \bar{z}$ are nowhere differentiable.

21. For $f(z)$ any one of the six trigonometric functions, verify that $\overline{f(z)} = f(\bar{z})$. Also deduce from Problems 18 to 20 that $f(\bar{z})$ is nowhere analytic.

22. Use the considerations of Section 16, in particular Theorem 4 and Eq. (38), to show directly from Eq. (41), taken as defining relations, that the functions $\cos z$ and $\sin z$ have derivatives given by Eq. (51) for every value of $z = x + iy$.

Verify that each u is harmonic, and use the relation of Problem 23 of Exercise 8, $f(z) = \int [u_x(z,0) - iu_y(z,0)]\, dz$, to show that

23. If $u = \cos x \cosh y$, $f(z) = \int \sin z\, dz = \cos z + Ci$.
24. If $u = \cos 3x \sinh 3y$, $f(z) = \int -3i \cos 3z\, dz = i \sin 3z + Ci$.
25. If $u = e^y \cos x$, $f(z) = \int (-\sin z + i \cos z)\, dz = e^{iz} + Ci$.

26. Show that the function $\tan z$ has the period π, and that $-\pi/2 < x \leq \pi/2$ is a *fundamental region*, in which each value $w \neq \pm i$ is assumed exactly once, while $\tan z$ is never equal to $\pm i$. *Hint:* From

$$w = \tan z = \frac{e^{iz} - e^{-iz}}{i(e^{iz} + e^{-iz})} = -i\frac{e^{2iz} - 1}{e^{2iz} + 1},$$

it follows that

$$e^{2iz} = -\frac{w - i}{w + i},\quad \text{if } w \neq \pm i.$$

24. The hyperbolic functions

The hyperbolic functions $\sinh z$ and $\cosh z$ are defined for complex z by the relations

$$\cosh z = \frac{e^z + e^{-z}}{2},\quad \sinh z = \frac{e^z - e^{-z}}{2} \qquad (53)$$

similar to Eq. (39). It follows from Eq. (4) that

$$\cosh z = \sum_{m=0}^{\infty} \frac{z^{2m}}{(2m)!}, \quad \sinh z = \sum_{m=0}^{\infty} \frac{z^{2m+1}}{(2m+1)!} \tag{54}$$

As these series converge everywhere, $\cosh z$ and $\sinh z$ are entire functions. We may deduce either from Eq. (53) or from Eq. (54) that

$$\frac{d(\cosh z)}{dz} = \sinh z, \quad \frac{d(\sinh z)}{dz} = \cosh z. \tag{55}$$

The remaining hyperbolic functions such as $\tanh z = \sinh z/\cosh z$ are defined by relations similar in form to Eq. (52).

We see from the series that $\cosh z$ is an *even* function, while $\sinh z$ is an *odd* function. That is

$$\cosh(-z) = \cosh z, \quad \sinh(-z) = -\sinh z. \tag{56}$$

It follows from Eq. (53) that

$$e^z = \cosh z + \sinh z, \quad e^{-z} = \cosh z - \sinh z. \tag{57}$$

We may use Eqs. (53) and (33) to prove that

$$\begin{aligned}\sin iz &= i \sinh z, \quad \cos iz = \cosh z, \\ \sinh iz &= i \sin z, \quad \cosh iz = \cos z.\end{aligned} \tag{58}$$

The first two are useful in obtaining identities involving hyperbolic functions from the corresponding trigonometric identities. For example, replacing z by iz in Eq. (44) leads to

$$\cosh^2 z - \sinh^2 z = 1. \tag{59}$$

Again, replacing z_1 by iz_1 and z_2 by iz_2 in Eq. (36) leads to

$$\begin{aligned}\cosh(z_1 + z_2) &= \cosh z_1 \cosh z_2 + \sinh z_1 \sinh z_2 \\ \sinh(z_1 + z_2) &= \sinh z_1 \cosh z_2 + \cosh z_1 \sinh z_2.\end{aligned} \tag{60}$$

It follows from Eqs. (60) and (58) that

$$\begin{aligned}\cosh(x + iy) &= \cosh x \cos y + i \sinh x \sin y \\ \sinh(x + iy) &= \sinh x \cos y + i \cosh x \sin y.\end{aligned} \tag{61}$$

Exercise 13

1. Prove that

$$|\cosh z|^2 = \sinh^2 x + \cos^2 y = \cosh^2 x - \sin^2 y.$$

Deduce from this that $\sinh |x| \leq |\cosh z| \leq \cosh x$.

Sec. 24 THE ELEMENTARY FUNCTIONS 63

2. Prove that
$$|\sinh z|^2 = \sinh^2 x + \sin^2 y = \cosh^2 x - \cos^2 y.$$
Deduce from this that $\sinh |x| \leq |\sinh z| \leq \cosh x$.

3. Prove that $\cosh z = 0$ if and only if $z = (\frac{1}{2} + k)\pi i$.

4. Prove that $\sinh z = 0$ if and only if $z = k\pi i$.

5. Prove that $\sinh(z + 2\pi i) = \sinh z$, $\cosh(z + 2\pi i) = \cosh z$, so that each of the functions $\sinh z$ and $\cosh z$ has the period $2\pi i$.

6. Use the considerations of Section 16, in particular Theorem 4 and Eq. (38), to show directly from Eq. (61), taken as defining relations, that the functions $\cosh z$ and $\sinh z$ have derivatives given by Eq. (55) for every value of $z = x + iy$.

Verify that each u is harmonic, and use the relation of Problem 23 of Exercise 8, $f(z) = \int [u_x(z,0) - iu_y(z,0)] \, dz$, to show that

7. If $u = \sinh x \sin y$,
$$f(z) = \int -i \sinh z \, dz = -i \cosh z + Ci.$$

8. If $u = \sinh 2x \cos 2y$,
$$f(z) = \int 2 \cosh 2z \, dz = \sinh 2z + Ci.$$

9. From Problem 12 of Exercise 12 and Eq. (58), or directly from Eq. (53), deduce that $-\pi < y \leq \pi$ is a *fundamental region* for each of the functions $\cosh z$ and $\sinh z$, in which $\cosh z$ takes on 1 and -1 at one point, $\sinh z$ takes on i and $-i$ at one point, and each function takes on every other value at exactly two points.

For the function $\tanh z = \dfrac{\sinh z}{\cosh z}$, deduce that

10. The function is analytic for every $z \neq (\frac{1}{2} + k)\pi i$.

11. $\dfrac{d(\tanh z)}{dz} = \dfrac{1}{\cosh^2 z} = \operatorname{sech}^2 z$, if $z \neq (\frac{1}{2} + k)\pi i$.

12. $\tanh(z + \pi i) = \tanh z$, so that $\tanh z$ has the period πi.

13. $-\pi/2 < y \leq \pi/2$ is a *fundamental region* in which each value of $w \neq \pm 1$ is assumed exactly once, while $\tanh z$ is never equal to ± 1. *Hint:* From
$$w = \tanh z = \frac{e^z - e^{-z}}{e^z + e^{-z}} = \frac{e^{2z} - 1}{e^{2z} + 1},$$
it follows that $e^{2z} = -\dfrac{w + 1}{w - 1}$, if $w \neq \pm 1$.

25. The logarithmic function

We define the logarithmic function as the inverse of the exponential function. That is,
$$w = \log z \quad \text{if} \quad z = e^w. \tag{62}$$
Since e^w is never zero, we must assume that $z \neq 0$. As in Section 7, let $z = r \operatorname{cis} \theta$. Then if $w = u + iv$, we have
$$re^{i\theta} = z = e^w = e^{u+iv} = e^u e^{iv}. \tag{63}$$
Taking absolute values, and using Eq. (15), we find that*
$$r = e^u, \quad u = \operatorname{Log} r. \tag{64}$$
Since $z \neq 0$, $r = |z| \neq 0$. And $e^{i\theta} = e^{iv}$ so that from Eq. (20),
$$iv = i\theta + 2k\pi i \quad \text{and} \quad v = \theta + 2k\pi. \tag{65}$$
Thus $w = u + iv = \operatorname{Log} r + i(\theta + 2k\pi)$. And, if $z \neq 0$,
$$\log z = \operatorname{Log} r + i(\theta + 2k\pi) \tag{66}$$
where $r = |z|$ and θ is any one value of arg z.

Since k may have any of the values $0, \pm 1, \pm 2, \ldots$, the function $\log z$ is *many-valued*. The values
$$\log z = \operatorname{Log} r + i\theta, \quad \text{with} \; -\pi < \theta \leq \pi, \tag{67}$$
which lie in the fundamental strip of Eq. (22), constitute the *principal branch*, and Eq. (67) defines the *principal* value of $\log z$. With the value of θ used in Eq. (67), each positive or negative integral value of k in Eq. (66) defines another *branch* of $\log z$.

Consider a particular point $z_0 \neq 0$. Let $|z_0| = r_0$ and θ_0 be any one value of arg z_0. Then by Eq. (66), one value of $\log z_0$ is $\operatorname{Log} r_0 + i\theta_0$. Let z vary continuously from z_0 to z along any curve not passing through the origin. Then $\log z = \operatorname{Log} |z| + i \arg z$ will be a continuous function along the curve if we take arg z as the value of the polar angle θ obtained from θ_0 by continuous variation. Suppose that z traverses the complete circle $|z| = r_0$. Then by this variation from z_0 back to z_0, the value of arg z will have increased or decreased by 2π, according as the circle is traversed in the positive or negative direc-

---

* In this section, to avoid using log with two meanings in the same equation, we write Log to mean the real natural logarithm of a positive number often denoted by ln in the calculus.

tion. This shows that the value of log z will have changed by $\pm 2\pi i$. Similarly, if z winds about the origin k times, the new value of the logarithm will be log $|z_0| + i\theta_0 + 2k\pi i$. Hence only the totality of values given by Eq. (66) can allow for all possible variations of $z \neq 0$.

In the domain obtained from the z-plane by removing all points of the negative real axis, any one branch given by Eq. (66) with a particular value of k, and with $-\pi < \theta < \pi$, is single-valued and continuous. Similarly, in the domain obtained from the z-plane by removing all points of the positive real axis, any one branch given by Eq. (66) with a particular value of k, and with $0 < \theta < 2\pi$, is single-valued and continuous. This second type of domain may be used for values of z which are real and negative. Any particular $z_0 \neq 0$ is an interior point of a domain of one of our two types. If θ_0 is any one value of arg z_0, there is a branch of log z, defined in this domain, for which log $z_0 =$ Log $|z_0| + i\theta_0$. To show that this branch is differentiable at z_0, we first prove a general theorem.

Theorem. Let w_0 and z_0 be corresponding values of the inverse functions $w = f(z)$, $z = F(w)$. At $w = w_0$, let $F(w)$ have a derivative $F'(w_0) \neq 0$, and let $f(z)$ be single-valued and continuous in some neighborhood of z_0. Then $f(z)$ has a derivative at z_0 whose value is

$$f'(z_0) = \frac{1}{F'(w_0)}. \tag{68}$$

Let z_1 be in the given neighborhood of z_0, and let $w_1 = f(z_1)$ so that $z_1 = F(w_1)$. Then from the assumed continuity of $w = f(z)$, $w_1 - w_0$ must approach zero when $z_1 - z_0$ approaches zero. The existence of $F'(w_0)$ implies that $F(w)$ is single-valued in some neighborhood of w_0. Hence for w_1 in this neighborhood, we must have $w_1 \neq w_0$ when $z_1 \neq z_0$. This shows that as $z_1 \to z_0$ through values $\neq z_0$, $w_1 \to w_0$ through values (eventually) $\neq w_0$. Hence we may conclude that

$$f'(z_0) = \lim_{z_1 \to z_0} \frac{f(z_1) - f(z_0)}{z_1 - z_0} = \lim_{w_1 \to w_0} \frac{w_1 - w_0}{F(w_1) - F(w_0)}$$
$$= \frac{1}{\lim_{w_1 \to w_0} \frac{F(w_1) - F(w_0)}{w_1 - w_0}} = \frac{1}{F'(w_0)}. \tag{69}$$

This proves the theorem.

THE ELEMENTARY FUNCTIONS

To apply the theorem to $\log z$, we take $F(w) = e^w$, and $f(z)$ as any branch of $\log z$ which is single-valued in a domain including $z_0 \neq 0$ as an interior point. Then $F'(w_0) = e^{w_0} \neq 0$, so that

$$\left[\frac{d(\log z)}{dz}\right]_{z_0} = f'(z_0) = \frac{1}{e^{w_0}} = \frac{1}{z_0}, \qquad (70)$$

since $z = F(w) = e^w$ implies that $e^{w_0} = z_0$. This shows that in any open region in which it is single-valued, every branch of $\log z$ is analytic. For any value of $z \neq 0$, we have

$$\frac{d(\log z)}{dz} = \frac{1}{z}. \qquad (71)$$

Let us next consider two values w_1 and w_2. And let us suppose that $z_1 = e^{w_1}$, $z_2 = e^{w_2}$. Then from Eqs. (6) and (7) we have

$$z_1 z_2 = e^{w_1 + w_2}, \quad \frac{z_1}{z_2} = e^{w_1 - w_2}. \qquad (72)$$

It follows from this and Eq. (62) that for *some* value of each logarithm,

$$\log z_1 z_2 = \log z_1 + \log z_2,$$
$$\log \frac{z_1}{z_2} = \log z_1 - \log z_2. \qquad (73)$$

Because the function $\log z$ is many-valued, these equations are true only in the sense that each value of one member is included in the possible values of the other member.

Exercise 14

1. Use the considerations of Problems 16 and 17 of Exercise 8 to show that $\text{Log } \sqrt{x^2 + y^2} = \text{Log } r$ is harmonic, and that in every open region where it is single-valued, $\tan^{-1}(y/x) = \theta$ is harmonic.

2. Use the considerations of Problems 5 and 6 of Exercise 7 to show directly from Eq. (66), taken as a defining relation, that $\log z$ has a derivative given by Eq. (71) for every $z \neq 0$.

From Problem 1 and the relation of Problem 23 of Exercise 8,

$$f(z) = \int [u_x(z,0) - iu_y(z,0)]\, dz,$$

deduce that

3. If $u = \frac{1}{2} \text{Log}(x^2 + y^2)$,
$$f(z) = \int \frac{dz}{z} = \log z + Ci.$$
4. If $u = \tan^{-1}(y/x)$,
$$f(z) = \int -i\frac{dz}{z} = -i \log z + Ci.$$

5. Let p denote any positive real number. Show that if $z = p$, the principal value of the logarithm given by Eq. (67) is Log p.

By Problem 5, we may designate the *principal value* of log z by Log z without contradicting our earlier interpretation of Log r. Verify each evaluation of the principal value Log z.

6. $\text{Log } 1 = 0$. 7. $\text{Log } i = \pi i/2$. 8. $\text{Log}(-i) = -\pi i/2$.

9. $\text{Log}(-1) = \pi i$. 10. $\text{Log}(1 + i\sqrt{3}) = \text{Log } 2 + \pi i/3$.

11. Let $F(w) = 1 + w + |w|^2/4$. Show that $F(w)$ has a derivative at $w_0 = 0$, $F'(0) = 1 \neq 0$. Also show that in the neighborhood of $z_0 = 1$, $|z - z_0| < 1$, the function $f(z) = u + iv$ with $u = -2 + \sqrt{4x - y^2}$, $v = y$ is single-valued and continuous. Deduce from the theorem of Eq. (68) that $f(z)$ has a derivative at $z_0 = 1, f'(1) = 1$.

12. Use Theorem 4 of Section 16 to check the result of Problem 11,
$$f'(1) = 1 \text{ if } f(z) = -2 + \sqrt{4x - y^2} + iy.$$

13. Show that for $|z| < 1$, for the principal value of the logarithm,
$$\text{Log}\left(\frac{1}{1-z}\right) = z + \frac{z^2}{2} + \ldots + \frac{z^n}{n} + \ldots.$$

Hint: Let $R(z) = \text{Log}\left(\frac{1}{1-z}\right) - \sum_{n=1}^{\infty} \frac{z^n}{n}$. Note that $R(0) = 0$, and deduce that
$$R'(z) = \frac{1}{1-z} - \sum_{n=0}^{\infty} z^n = 0, \text{ for } |z| < 1,$$
so that $R(z) = 0$.

14. Prove that for the principal value of the logarithm, if $|Z - 1| \leq p < 1$, $|\text{Log } Z| \leq p/(1-p)$. *Hint:* From Problem 13, with $z = 1 - Z$,
$$\text{Log } Z = \sum_{n=1}^{\infty} \frac{1}{n}(-1)^{n+1}(Z-1)^n \text{ if } |Z - 1| < 1.$$

Hence if $|Z - 1| \leq p < 1$,

$$|\text{Log } Z| \leq \sum_{n=1}^{\infty} \frac{p^n}{n} \leq \sum_{n=1}^{\infty} p^n = \frac{p}{1-p}.$$

Since $p^n/n < p^n$ if $p \neq 0$, $|\text{Log } Z| < p/(1-p)$ unless $p = 0$, $Z = 1$.

26. The general power function

Let A be any complex number. Then for $z \neq 0$, we define

$$z^A = e^{A \log z}. \tag{74}$$

This determines a value of the power for each choice of a particular branch of the function $\log z$. From Eq. (66) we have

$$z^A = e^{A[\text{Log } r + i(\theta + 2k\pi)]} = e^{A(\text{Log } r + i\theta)} e^{2Ak\pi i}. \tag{75}$$

The value of the power for a branch with $k \neq 0$ will agree with that for $k = 0$ only if

$$e^{2Ak\pi i} = 1 = e^0 \quad \text{or} \quad 2Ak\pi i = 0 + 2K\pi i, \tag{76}$$

by Eq. (20). Hence $Ak = K$ and $A = K/k$, a rational number.

If $A = 0$, $z^0 = 1$ for $z \neq 0$ and any k, by Eq. (75). If $A = n$, a positive (or negative) integer, nk is an integer for all values of k. Hence the last factor in Eq. (75) is unity, and from Eq. (75),

$$z^n = e^{n(\text{Log } r + i\theta)} = r^n e^{in\theta} = r^n \text{ cis } n\theta, \tag{77}$$

in agreement with Eqs. (47) and (48) of Section 9. Thus for n integral the definition gives a unique value in agreement with that based on repeated multiplication of z by itself.

Again, for m integral, and a particular determination of $\log z$,

$$z^{Am} = e^{Am \log z} = (e^{A \log z})^m = (z^A)^m. \tag{78}$$

Thus a rational power m/n is uniquely determined from the value of the power $1/n$, and we may take n positive. But from Eq. (75),

$$z^{1/n} = e^{1/n[\text{Log } r + i(\theta + 2k\pi)]} = r^{1/n} \text{ cis } \frac{\theta + 2k\pi}{n} \tag{79}$$

This, like Eq. (51) of Section 9, gives n distinct nth roots if we take $k = 0, 1, 2, \ldots, n - 1$. Since any other integer differs from one of these by a multiple of n, the definition gives n possible values to $z^{1/n}$.

These are the n roots of the equation $w^n - z = 0$, often denoted by $\sqrt[n]{z}$ or $\sqrt{z}$ for $z^{1/2}$.

When A is not real and rational, Eq. (75) will lead to different values for any two distinct values of k. Thus z^A is infinitely many-valued. But in any region in which $\log z$ is single-valued and continuous, there is a single-valued branch of z^A.

For any complex number $B \neq 0$, let $b = |B|$, and β be one value of arg B. Then from Eq. (75) with A replaced by z, we have

$$B^z = e^{z[\log b + i(\beta + 2k\pi)]}.\tag{80}$$

Each choice of k determines a separate function, single-valued and continuous for all complex values of z.

For B real and positive, $B = b$. Hence if we choose $\beta = 0$ and $k = 0$ in Eq. (80), we obtain the particular function

$$b^z = e^{z \log b}.\tag{81}$$

We usually take this as the definition of the exponential function to an arbitrary positive real base. For $b = e$ this is consistent with Eq. (4). The possibility of using b^z to mean any one of the infinitely many functions found by putting $B = b$ in Eq. (80) is occasionally convenient, as for example, when evaluating a particular branch of z^A for $z = b$ or $z = e$.

27. Inverse functions

The inverse trigonometric and inverse hyperbolic functions are all defined by relations like

$$w = \cos^{-1} z \quad \text{if} \quad z = \cos w.\tag{82}$$

As in Problem 10 of Exercise 12, we may deduce from this that when

$$w = \cos^{-1} z, \quad e^{iw} = z \pm \sqrt{z^2 - 1},\tag{83}$$

and as in Problem 11 and the hint to Problem 26 of Exercise 12, we have when

$$w = \sin^{-1} z, \quad e^{iw} = iz \pm \sqrt{1 - z^2},\tag{84}$$

$$w = \tan^{-1} z, \quad e^{-2iw} = \frac{i + z}{i - z}.\tag{85}$$

By taking logarithms and solving for w we may deduce that:

$$\cos^{-1} z = \sec^{-1} \frac{1}{z} = -i \log (z + \sqrt{z^2 - 1}), \qquad (86)$$

$$\sin^{-1} z = \csc^{-1} \frac{1}{z} = -i \log (iz + \sqrt{1 - z^2}), \qquad (87)$$

$$\tan^{-1} z = \cot^{-1} \frac{1}{z} = \frac{i}{2} \log \frac{i+z}{i-z} = \frac{1}{2i} \log \frac{1+zi}{1-zi}. \qquad (88)$$

We have omitted the $\pm$ sign as unnecessary, since by Eq. (79), $\sqrt{z} = z^{1/2}$ indicates either one of the two square roots.

A similar procedure leads to the expressions

$$\cosh^{-1} z = \operatorname{sech}^{-1} \frac{1}{z} = \log (z + \sqrt{z^2 - 1}), \qquad (89)$$

$$\sinh^{-1} z = \operatorname{csch}^{-1} \frac{1}{z} = \log (z + \sqrt{1 + z^2}), \qquad (90)$$

$$\tanh^{-1} z = \coth^{-1} \frac{1}{z} = \frac{1}{2} \log \frac{1+z}{1-z}. \qquad (91)$$

Each of these inverse functions is many-valued. But for each function there are branches which are single-valued and continuous in appropriate regions. That each branch is analytic for suitably restricted values of z follows either from the chain rule for composite functions of Section 15, or from the theorem of Eq. (68). The latter argument leads more directly to the differentiation formulas, as illustrated in Problems 8 to 13 of Exercise 15.

28. Elementary functions

For complex values of the variable, we consider the exponential and logarithmic functions e^z and $\log z$ as the two *basic elementary* functions. And by the four *fundamental operations* we mean addition, subtraction, multiplication, and division. We then define:

An *elementary function* of the complex variable z is a function which can be explicitly represented in terms of complex constants and the independent variable z by means of the four fundamental operations and the two basic functions, using, at most, a finite number of operations, and at most, a finite number of basic functions. Usually,

By taking logarithms and solving for w we may deduce that:

$$\cos^{-1} z = \sec^{-1} \frac{1}{z} = -i \log (z + \sqrt{z^2 - 1}), \qquad (86)$$

$$\sin^{-1} z = \csc^{-1} \frac{1}{z} = -i \log (iz + \sqrt{1 - z^2}), \qquad (87)$$

$$\tan^{-1} z = \cot^{-1} \frac{1}{z} = \frac{i}{2} \log \frac{i + z}{i - z} = \frac{1}{2i} \log \frac{1 + zi}{1 - zi}. \qquad (88)$$

We have omitted the $\pm$ sign as unnecessary, since by Eq. (79), $\sqrt{z} = z^{1/2}$ indicates either one of the two square roots.

A similar procedure leads to the expressions

$$\cosh^{-1} z = \operatorname{sech}^{-1} \frac{1}{z} = \log (z + \sqrt{z^2 - 1}), \qquad (89)$$

$$\sinh^{-1} z = \operatorname{csch}^{-1} \frac{1}{z} = \log (z + \sqrt{1 + z^2}), \qquad (90)$$

$$\tanh^{-1} z = \coth^{-1} \frac{1}{z} = \frac{1}{2} \log \frac{1 + z}{1 - z}. \qquad (91)$$

Each of these inverse functions is many-valued. But for each function there are branches which are single-valued and continuous in appropriate regions. That each branch is analytic for suitably restricted values of z follows either from the chain rule for composite functions of Section 15, or from the theorem of Eq. (68). The latter argument leads more directly to the differentiation formulas, as illustrated in Problems 8 to 13 of Exercise 15.

28. Elementary functions

For complex values of the variable, we consider the exponential and logarithmic functions e^z and $\log z$ as the two *basic elementary* functions. And by the four *fundamental operations* we mean addition, subtraction, multiplication, and division. We then define:

An *elementary function* of the complex variable z is a function which can be explicitly represented in terms of complex constants and the independent variable z by means of the four fundamental operations and the two basic functions, using, at most, a finite number of operations, and at most, a finite number of basic functions. Usually, for any choice of branches of the logarithmic functions involved, we may find two-dimensional regions in which the representation leads to a single-valued and continuous branch, whose derivative can be found by the rules of Section 15. We call such an analytic branch a *single-valued elementary function*. Polynomials and rational functions are elementary. That z^A, and in particular $\sqrt[N]{z}$, are elementary follows from Eqs. (74) and (79). That $\cos z$ and $\sin z$ are elementary follows from Eq. (33), while that $\cosh z$ and $\sinh z$ are elementary follows from Eq. (53). By Eq. (52) and the similar relations, the remaining trigonometric and hyperbolic functions are elementary. The nature of the definition of elementary function is such that any finite combination of elementary functions is again elementary. Hence it follows from Section 27 that each of the inverse trigonometric and inverse hyperbolic functions is an elementary function.

We note that the inverse of an elementary function is not necessarily elementary. For example, the function $w = f^{-1}(z)$ if $z = f(w)$ is not elementary if $z = w + e^w$, or if $z = w + w^5$.

29. Riemann surfaces

We have seen that several elementary functions, such as $z^{1/n}$, $\log z$, $\sin^{-1} z$ are many-valued. Let us in particular consider the function $w = \sqrt{z} = z^{1/2}$. By Eq. (79), all the values of this function are given by

$$w_1 = \sqrt{r} \operatorname{cis} \frac{\theta}{2}, \quad -\pi < \theta \leq \pi,$$
$$w_2 = \sqrt{r} \operatorname{cis} \left(\frac{\theta}{2} + \pi\right), \quad -\pi < \theta \leq \pi. \tag{92}$$

Since $\theta = \pi$ and $\theta = -\pi$ are each the equation of the negative real axis in the z-plane, the condition $-\pi < \theta \leq \pi$ prevents the point z from crossing the negative axis. We may regard the domain $-\pi < \theta \leq \pi$ in which each branch is single-valued as obtained from the z-plane by the following construction. Make a *cut* in the plane from 0 to ∞ along the *negative* real axis. The upper edge, or bank, of the cut belongs to the cut plane since θ may equal π. The lower edge, or bank, of the cut does not belong to the cut plane since θ may not equal $-\pi$.

We now consider two replicas of the cut plane, P_1 and P_2. We think of them as superimposed, and we make the convention that the upper edge of the cut in P_1 is joined to the lower edge of the cut in P_2, while the upper edge of the cut in P_2 is joined to the lower edge of the cut in P_1. Suppose now that the moving point z traverses a circle with center at the origin in the counterclockwise direction. If we start with $\theta = 0$ for $z = r$, and $w = w_1 = \sqrt{r}$, when halfway round z will be r cis π and $w = w_1 = \sqrt{r}$ cis $(\pi/2) = \sqrt{r}i$, corresponding to the upper edge in P_1. Let arg z increase beyond π to $\pi + h$. Then arg $z = \theta + 2\pi$, since $\theta = -\pi + h$. Thus w will change continuously to $\sqrt{r}$ cis $(\pi/2 + h/2) = \sqrt{r}$ cis $(\theta/2 + \pi) = w_2$, corresponding to a point near the lower edge in P_2. If we return to $z = r$ cis 2π, and go around a second time, when halfway round z will be r cis 3π and $w = w_2 = \sqrt{r}$ cis $(3\pi/2) = -ri$, corresponding to the upper edge in P_2. Let arg z increase beyond 3π to $3\pi + h$. Then arg $z = \theta + 4\pi$, since $\theta = -\pi + h$. Thus w will change continuously to become $\sqrt{r}$ cis $(3\pi/2 + h/2) = \sqrt{r}$ cis $(\theta/2 + 2\pi) = \sqrt{r}$ cis $\theta = w_1$, corresponding to a point near the lower edge in P_1. The two superimposed planes constitute the Riemann surface for $w = \sqrt{z}$. The discussion just given shows that although the function $w = \sqrt{z}$ is two-valued in the z-plane for $z \neq 0$, the function is a single-valued and continuous function of position on the two-sheeted Riemann surface.

The cuts are to some extent arbitrary, and we might have cut the plane along any straight (or curved) line extending from the origin to infinity. But the point $z = 0$ has a special relation to the function. Let z describe a small circle with center $z_0 \neq 0$ and radius less than $|z_0|$, so that the origin lies outside the circle. Then arg z is not changed by the circuit, and the final value of w will be the same as the initial value, but if z describes any circle which includes the origin in its interior, arg z will change by 2π, and the values of w_1 and w_2 will be interchanged by the circuit. If a circuit about a point can change the value of a many-valued function, the point is a *branch point*. Thus $z = 0$ is a branch point of the function $w = \sqrt{z}$. As we shall study more fully in Section 60, a large circuit about any point is considered

Sec. 29   THE ELEMENTARY FUNCTIONS   73

to be a circuit about $z = \infty$, and the point $z = \infty$ is also a branch point of $w = \sqrt{z}$.

For the function $w = \sqrt{z - a}$, the branch points are $z = a$ and $z = \infty$. Here the cut may be taken as any half line from a to infinity.

The relation $w = \sqrt[n]{z} = z^{1/n}$ has n branches. These may be given by Eq. (79) with $-\pi < \theta \leq \pi$, $k = 0, 1, 2, \ldots, n - 1$. The branch points are $z = 0$ and $z = \infty$, and for these branches the cut is again the negative real axis. Here we must use n replicas of the cut plane P_n corresponding to the branch with $k = n - 1$, and we must join the upper edge of the cut in P_n to the lower edge in P_1, and the upper edge in P_m to the lower edge in P_{m+1} for $m < n$.

For $w = \log z$ there are an infinite number of branches. These may be given by Eq. (66) with $-\pi < \theta \leq \pi$. The branch points are $z = 0$ and $z = \infty$, and for these branches the cut is again the negative real axis. Here we must use an infinite number of replicas of the cut plane P_k corresponding to the branch of Eq. (66), where $k = 0, \pm 1, \pm 2, \ldots$, and we must join the upper edge of the cut in P_k to the lower edge in P_{k+1}. The only two boundary points of this infinitely many-sheeted Riemann surface are the two branch points $z = 0$ and $z = \infty$. All other points are interior to the surface.

For $w = \sqrt{(z - a_1)(z - a_2)}$, $a_1 \neq a_2$, there are two sheets and two branch points $z = a_1$, $z = a_2$. The cut may be taken as the straight-line segment joining a_1 and a_2. As for $w = \sqrt{z}$, the replicas P_1 and P_2 are joined crosswise along the cut. Note that in this case $z = \infty$ is *not* a branch point.

For $w = \sqrt{(z - a_1)(z - a_2)\ldots(z - a_n)}$, with all the a_m distinct, there are two sheets. Each a_m is a branch point, and $z = \infty$ is a branch point if n is odd, but *not* if n is *even*. The cuts may be taken as the straight-line segments joining a_1 to a_2, a_3 to a_4, and so on with an additional half line from a_n to ∞ if n is odd. As for $w = \sqrt{z}$, the replicas P_1 and P_2 are joined crosswise along each of the cuts.

Exercise 15

1. Show that, for any $z \neq 0$, $d(z^A)/dz = Az^{A-1}$, where the same branch of $\log z$ is used to find z^A and z^{A-1}.

THE ELEMENTARY FUNCTIONS

2. Show that for $|z| < 1$, the "principal" branch of
$$(1 + z)^A = e^{A \, \text{Log} \, (1+z)}$$
$$= 1 + \sum_{n=1}^{\infty} \frac{A(A-1)\ldots(A-n+1)}{n!} z^n.$$

Hint: Let $S(z)$ be the sum function of the series plus unity, or right-hand member. Deduce that $(1+z)S'(z) = AS(z)$. Hence if $R(z) = \text{Log} \, S(z) - A \, \text{Log} \, (1+z)$, for $|z| < 1$, $R'(z) = 0$, so that $R(z) = R(0) = 0$.

Verify each evaluation of the general power:

3. $1^{\sqrt{2}} = \text{cis} \, 2\sqrt{2} k\pi$.   **4.** $i^i = e^{(-\pi/2) - 2k\pi}$.

5. Verify that $z^A z^B$ usually takes on more values than z^{A+B}. *Hint:* $\log(z^A z^B) = (A+B) \, \text{Log} \, z + (Ak_1 + Bk_2)2\pi i$, which is a value of $\log(z^{A+B})$ only if $Ak_1 + Bk_2 = (A+B)k_3 + k_4$.

6. Verify that $(z^A)^B$ usually takes on more values than z^{AB}. *Hint:* $\log(z^A)^B = AB \, \text{Log} \, z + (ABk_1 + Bk_2)2\pi i$, which is a value of $\log z^{AB}$ only if $ABk_2 + Bk_2 = ABk_3 + k_4$.

7. Show that for $B \neq 0$, $d(B^z)/dz = B^z \log B$, where $\log B$ is on the branch which makes $B^z = e^{z \log B}$.

Prove each of the following differentiation rules by an argument similar to that used in the text to derive Eq. (71). Note that where a square root appears, we must use that one of its two values which satisfies the supplementary condition.

8. $\dfrac{d(\sin^{-1} z)}{dz} = \dfrac{1}{\sqrt{1-z^2}}$, $z \neq \pm 1$, $\sqrt{1-z^2} = \cos(\sin^{-1} z)$.

9. $\dfrac{d(\cos^{-1} z)}{dz} = \dfrac{-1}{\sqrt{1-z^2}}$, $z \neq \pm 1$, $\sqrt{1-z^2} = \sin(\cos^{-1} z)$.

10. $\dfrac{d(\tan^{-1} z)}{dz} = \dfrac{1}{1+z^2}$, $z \neq \pm i$.

11. $\dfrac{d(\sinh^{-1} z)}{dz} = \dfrac{1}{\sqrt{1+z^2}}$, $z \neq \pm i$, $\sqrt{1+z^2} = \cosh(\sinh^{-1} z)$.

12. $\dfrac{d(\cosh^{-1} z)}{dz} = \dfrac{1}{\sqrt{z^2-1}}$, $z \neq \pm 1$, $\sqrt{z^2-1} = \sinh(\cosh^{-1} z)$.

13. $\dfrac{d(\tanh^{-1} z)}{dz} = \dfrac{1}{1-z^2}$, $z \neq \pm 1$.

14. Check Problem 10 by using Eq. (88).
15. Check Problem 13 by using Eq. (91).

In Problem 2 replace z by z^2 and deduce that for $|z| < 1$,

16. $\dfrac{1}{1 + z^2} = 1 - z^2 + z^4 - \ldots + (-1)^m z^{2m} + \ldots$

17. $\dfrac{1}{\sqrt{1 - z^2}} = 1 + \dfrac{1}{2}z^2 + \dfrac{1\cdot 3}{2\cdot 4}z^4 + \dfrac{1\cdot 3\cdot 5}{2\cdot 4\cdot 6}z^6 + \ldots$

18. From Problems 8 and 17 deduce that for $|z| < 1$, and the branch of $\sin^{-1} z$ with $\sin^{-1} 0 = 0$ (which makes $\sin^{-1} z$ real for z real),

$$\sin^{-1} z = z + \frac{1}{2}\frac{z^3}{3} + \frac{1\cdot 3}{2\cdot 4}\frac{z^5}{5} + \frac{1\cdot 3\cdot 5}{2\cdot 4\cdot 6}\frac{z^7}{7} + \ldots$$

19. From Problems 10 and 16 deduce that for $|z| < 1$, and the branch of $\tan^{-1} z$ with $\tan^{-1} 0 = 0$ (which makes $\tan^{-1} z$ real for z real),

$$\tan^{-1} z = z - \frac{z^3}{3} + \frac{z^5}{5} \cdots + (-1)^m \frac{z^{2m+1}}{2m+1} + \ldots$$

20. Check Problem 19 by using Eq. (88) and Problem 13 of Exercise 14. Hence deduce that the branch of Problem 19 has $-\pi/2 < \operatorname{Re} \tan^{-1} z < \pi/2$. Compare Problem 26 of Exercise 12.

21. From Problem 18, deduce that for $|z| < 1$, $|\sin^{-1} z| \leq \sin^{-1} |z| < \pi/2$. Hence in particular, the branch of Problem 18 has $-\pi/2 < \operatorname{Re} \sin^{-1} z < \pi/2$. Compare Problems 12 and 15 of Exercise 12.

22. For $w = \sqrt{1 + z^2}$, let z describe a path OA which, except for its end points, lies interior to the first quadrant. Show that if $w = 1$ at $z = 0$, for A on the positive real axis w will be real and positive, but for A on the positive imaginary axis, w will be real and positive for $\operatorname{Im} z < 1$ and be a positive multiple of i for $\operatorname{Im} z > 1$.

23. From Eq. (88) deduce that the function $\tan^{-1} z$ has single-valued branches

$$\tan^{-1} z = \frac{1}{2i}[\operatorname{Log}(1 + zi) - \operatorname{Log}(1 - zi)] + k\pi$$

in the cut plane obtained by excluding all points of the two half lines which extend from $\pm i$ to infinity outward along the imaginary axis, $\operatorname{Re} z = 0$. The principal branch with $k = 0$ has $-\pi/2 <$

Re $\tan^{-1} z < \pi/2$. The branch points are at $\pm i$. And $z = \infty$ is *not* a branch point.

24. From Eq. (87) deduce that the function $\sin^{-1} z$ has single-valued branches

$$\sin^{-1} z = 2m\pi - i \operatorname{Log}(iz + \sqrt{1 - z^2}),$$

$k = 2m$, even, and

$$\sin^{-1} z = (2m + 1)\pi + i \operatorname{Log}(iz + \sqrt{1 - z^2}),$$

$k = 2m + 1$, odd, in the cut plane obtained by excluding all the points of the two half lines which extend outward along the real axis from ± 1 to infinity, $\operatorname{Im} z = 0$, $\operatorname{Re} z < -1$ and $\operatorname{Re} z > 1$. Let $\sqrt{1 - z^2}$ mean the branch of the square root in the cut plane with positive real part. Then from

$$(iz + \sqrt{1 - z^2})(iz - \sqrt{1 - z^2}) = -1,$$

we may deduce that

$$\operatorname{Log}(iz + \sqrt{1 - z^2}) + \operatorname{Log}(iz - \sqrt{1 - z^2}) = i\pi$$

for the values approached on the cut through $+1$, but $= -i\pi$ for the values approached on the cut through -1. It follows that we must join P_{2m} to P_{2m-1} crosswise along the positive cut, and P_{2m} to P_{2m+1} crosswise along the negative cut. The points $z = \pm 1$, $z = \infty$ are the branch points. The principal branch is obtained by taking $k = 2m = 0$, and assigning the special meaning to $\sqrt{1 - z^2}$.

5

CONFORMAL TRANSFORMATIONS

A functional relation between two complex variables $w = f(z)$ sets up a correspondence between points (x,y), where $z = x + iy$, and other points (u,v), where $w = u + iv$. If we think of two separate complex planes, we may visualize a mapping of portions of the z-plane upon portions of the w-plane. The mapping of corresponding curves and regions in the two planes is often helpful in studying properties of the function. Sometimes we identify the w-plane with the z-plane and imagine a mapping or transformation of this plane onto itself. This permits a graphic description of many simple transformations such as translations and rotations. We first require that $u(x,y)$ and $v(x,y)$ have continuous partial derivatives, and discuss an approximating linear transformation. Then we require the function $f(z)$ to be an analytic function of z, and show that this leads to mappings which have a conformal, or angle-preserving property. Finally we give some examples of the usefulness of the conformal mapping of regions in simplifying the solution of certain physical problems, such as that of Dirichlet.

30. Functions as transformations

Let $w = u + iv$ be a single-valued function of the complex variable $z = x + iy$, as in Section 12. Thus for all z in some set S, we have

$$w = f(z) = u(x,y) + iv(x,y). \tag{1}$$

We may regard this as a *transformation* of certain points of the complex plane, which takes each point (x,y) into a corresponding point (u,v).

It is often convenient to use *two* separate complex planes, a z-plane and a w-plane, Fig. 11. In the first plane we plot the point $z = (x,y)$. In the second plane we plot the corresponding point $w = (u,v)$. In this way, an *image* point w is associated with each point z in S, so that S is *mapped* on the w-plane.

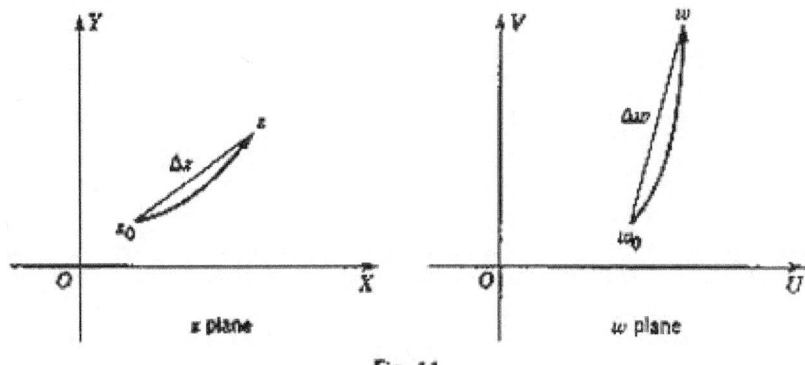

Fig. 11

The mapping is continuous when $f(z)$ is continuous, as defined in Section 13. In this case, for any fixed z_0 and given $\epsilon > 0$, there is a $\delta(\epsilon)$ for which

$$|\Delta w| = |w - w_0| < \epsilon \quad \text{if} \quad |\Delta z| = |z - z_0| < \delta(\epsilon). \tag{2}$$

Thus changes in w are small for sufficiently small changes in z. In this sense neighboring points in the z-plane correspond to neighboring points in the w-plane. Consequently, under continuous mapping, the image of a continuous curve is in general also a continuous curve, although it may degenerate to a point.

31. Continuously differentiable transformations

Suppose that $u(x,y)$ and $v(x,y)$ each have continuous first partial derivatives. Then by Eqs. (31) and (32) of Section 16, we may write

$$\Delta u = \Delta x(u_x + \epsilon_1) + \Delta y(u_y + \epsilon_2),$$
$$\Delta v = \Delta x(v_x + \epsilon_3) + \Delta y(v_y + \epsilon_4), \tag{3}$$

where the ϵ's all approach zero when Δx and Δy both approach zero. This suggests the linear equations

$$u - u_0 = (x - x_0)u_x(x_0,y_0) + (y - y_0)u_y(x_0,y_0),$$
$$v - v_0 = (x - x_0)v_x(x_0,y_0) + (y - y_0)v_y(x_0,y_0), \tag{4}$$

obtained by putting all the ϵ's equal to zero in Eq. (3), as a possible approximation to $u = u(x,y)$, $v = v(x,y)$ for points x,y near x_0,y_0. The possibility of solving these equations for x and y in terms of u and v depends on the nature of J at x_0,y_0 where

$$J = \begin{vmatrix} u_x & u_y \\ v_x & v_y \end{vmatrix} = u_x v_y - u_y v_x \tag{5}$$

is the Jacobian (determinant) of the two functions $u(x,y)$ and $v(x,y)$.

If $J = 0$ at the point x_0,y_0, there is no unique solution. The mapping given by Eq. (4) degenerates, taking the whole z-plane into a single line or point. In any neighborhood of u_0,v_0 there may be points u,v for which the equations $u = u(x,y)$, $v = v(x,y)$ lead to none, one, or many values of x and y. When $J = 0$ at the point x_0,y_0, then Eq. (4) is of little help in understanding the nature of the transformation $u = u(x,y)$, $v = v(x,y)$. See Problems 1 to 7 of Exercise 16.

If $J \neq 0$ at the point x_0,y_0, the simultaneous system of Eq. (4) has a unique solution, and one can find x and y explicitly as first-degree expressions in u and v. Hence the mapping given by Eq. (4) is one-to-one everywhere, and maps the entire z-plane on the entire w-plane. Such first-degree equations take parallel lines into parallel lines. Equally spaced points on any one line in one plane go into equally spaced points on some line in the second plane. However, while the scale factor or ratio of the length of a segment to that of its image segment is the same for any two parallel lines, it will usually differ for different directions. Concentric circles about x_0,y_0 will go into a set of concentric, similar, and similarly oriented ellipses with center at u_0,v_0, Fig. 12. When $J \neq 0$ at x_0,y_0, the transformation given by Eq. (4) does approximate that for $u = u(x,y)$, $v = v(x,y)$ to within any fixed positive fraction of $\sqrt{u^2 + v^2}$, as shown in Problem 11 of Exercise 16.

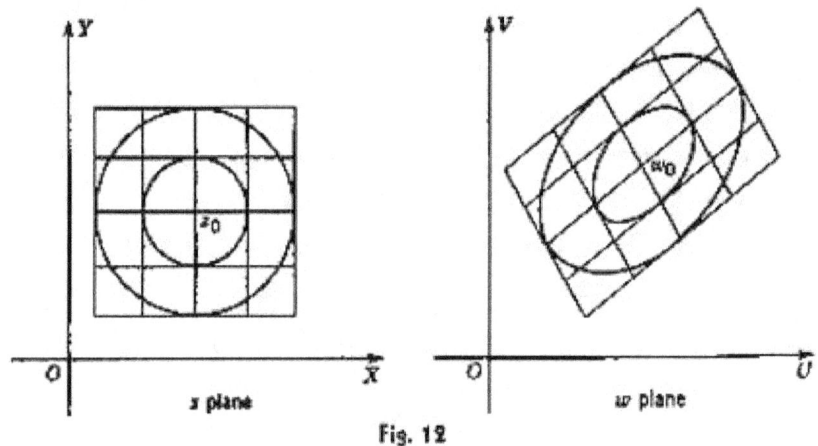

Fig. 12

That the transformation $u = u(x,y)$, $v = v(x,y)$ is itself one-one in some neighborhood of x_0,y_0 when $J \neq 0$ follows from the theorem on *implicit functions* of two real variables which states that:

Let the functions $u = u(x,y)$, $v = (x,y)$ *each possess continuous first partial derivatives with respect to* x *and* y *for the particular value* x_0,y_0. *Let* $u_0 = u(x_0,y_0)$ *and* $v_0 = v(x_0,y_0)$. *Then if the Jacobian* $J = u_x v_y - u_y v_x \neq 0$ *at* x_0,y_0, *the two equations* $u = u(x,y)$, $v = v(x,y)$ *may be simultaneously satisfied by two functions* $x = x(u,v)$, $y = y(u,v)$ *such that* $x_0 = x(u_0,v_0)$, $y_0 = y(u_0,v_0)$. *These functions are uniquely determined and continuous in some two-dimensional neighborhood of* x_0,y_0, *and in this region possess continuous first partial derivatives such that the equations*

$$1 = x_u u_x + x_v v_x, \qquad 0 = y_u u_x + y_v v_x,$$
$$0 = x_u u_y + x_v v_y, \qquad 1 = y_u u_y + y_v v_y, \qquad (6)$$

obtained by partial differentiation of $x = x(u,v)$, $y = y(u,v)$ *are satisfied.*

For a proof, we refer the reader to Chapter 10 of the author's *A Treatise on Advanced Calculus.*

Sec. 31   CONFORMAL TRANSFORMATIONS   81

As shown in Problem 12 of Exercise 16, Eq. (6) is the condition that the approximating linear equations for the inverse transformation at u_0, v_0 are the equations which solve the simultaneous system of Eq. (4).

We may briefly summarize the result of our discussion by observing that *for any point z_0 where the Jacobian $J = u_x v_y - u_y v_x \neq 0$, there is a neighborhood of this point in which the continuously differentiable mapping $u = u(x,y), v = v(x,y)$ is one-one, and is approximated by the linear transformation of Eq. (4) to within any preassigned fraction of $|w - w_0|$.*

Exercise 16

For each of the following pairs of mapping functions verify that if $z_0 = 0$, $w_0 = 0$ and that the Jacobian of Eq. (5) is zero at $(0,0)$. Also verify the statements about the inverse transformation, and the linear "approximation" of Eq. (4), for values near z_0.

1. $u = x$, $v = 0$, equal to its "approximation" has as inverse $x = u, y$ arbitrary for $v = 0$. The inverse is infinitely many-valued for $v = 0$, but there is no inverse for $v \neq 0$.

2. $u = 0, v = 0$, equal to its "approximation" has no inverse unless $u = 0, v = 0$. For $u = 0, v = 0$ the infinitely many-valued inverse is any point (x,y).

3. $u = x, v = y^3$ has a unique inverse $x = u$, $y = \sqrt[3]{v}$, and is one-one. The "approximation" is $u = x, v = 0$.

4. $u = x(x^2 + 4y^2)$, $v = y(x^2 + 4y^2)$ has a unique inverse, $z_0 = 0$ for $w_0 = 0$ and for $w_0 \neq 0$, $x = \dfrac{u}{\sqrt[3]{u^2 + 4v^2}}$, $y = \dfrac{v}{\sqrt[3]{u^2 + 4v^2}}$, with $\sqrt[3]{u^2 + 4v^2} = x^2 + 4y^2 > 0$. The mapping is one-one. The "approximation" is $u = 0, v = 0$.

5. $w_0 = 0$ for $z_0 = 0$ and for $z_0 \neq 0$, $u = \dfrac{x(x^2 + y^2)}{\sqrt{x^2 + 4y^2}}, v = \dfrac{2y(x^2 + y^2)}{\sqrt{x^2 + 4y^2}}$, has a unique inverse $x = \dfrac{2u\sqrt[4]{u^2 + v^2}}{\sqrt{4u^2 + v^2}}$, $y = \dfrac{v\sqrt[4]{u^2 + v^2}}{\sqrt{4u^2 + v^2}}$ for $w_0 \neq 0$, and $z_0 = 0$ for $w_0 = 0$. The mapping is one-one and takes concentric circles about z_0 into circles about w_0, since $u^2 + v^2 = (x^2 + y^2)^2$. The "approximation" is $u = 0, v = 0$.

6. $u = x^2, y = x^2$ has as inverse $x = \pm\sqrt{u}, y = \sqrt{v}$ or $x = \pm\sqrt{u}$, $y = -\sqrt{v}$. The inverse is four-valued if $u > 0, v > 0$, two-valued if $u = 0, v > 0$, or $u > 0, v = 0$, one-valued if $u = 0, v = 0$, and has no (real) values if u or v is negative. The "approximation" is $u = 0$, $v = 0$.

7. $u = 0$ when $x = 0$, $u = x^4 \sin(1/x)$ for $x \neq 0$. $v = y$. The inverse $y = v$, x a root of $x^4 \sin(1/x) = u$ is many-valued for small $u \neq 0$, the number of roots increasing as $u \to 0$, to the infinite number given by $0, 1/k\pi$ for $u = 0$. The "approximation" is $u = 0, v = 0$.

8. Except for changes of origin in both planes, Eq. (4) has the form $u = ax + by$, $v = cx + dy$. Let $ad - bc = D$. Then $Dx = ud - bv$, $Dy = -cu + av$. Verify that if $x^2 + y^2 = r^2$, then $(c^2 + d^2)u^2 - 2(ac + bd)uv + (a^2 + b^2)v^2 = D^2r^2$. When $J \neq 0$, $D = J \neq 0$, and this represents an ellipse as in Fig. 12.

9. Under a rotation of coordinate axes, for the ellipse $Ax^2 + 2Bxy + Cy^2 = 1$, the discriminant $(2B)^2 - 4AC$ is invariant, and hence $= -\dfrac{4}{R_1^2 R_2^2}$, where R_1 and R_2 are the semiaxes, since the equation of the ellipse is $\dfrac{x^2}{R_1^2} + \dfrac{y^2}{R_2^2} = 1$ when referred to principal axes. Use this fact to show that for the ellipse of Problem 8, $\dfrac{1}{R_1^2 R_2^2} = \dfrac{1}{D^2 r^4}$. Hence $\pi R_1 R_2 = |D|\pi r^2$. Check this relation by an interpretation in terms of areas, and of the here constant Jacobian $J = D$ as the ratio of areas.

10. The relation of Problem 9, $R_1 R_2 = |D|r^2$ shows that $R_1 \neq 0$, where $R_1 < R_2$. And the equation of Problem 8 shows that $R_1/r = m_1$ and $R_2/r = m_2$ are independent of the size of r. But, if R is any radius $\sqrt{u^2 + v^2}$, from $R_1 \leq R \leq R_2$, we may deduce that $m_1 \leq R/r \leq m_2$, and in particular, $R/r \geq m_1 > 0$ or $r \leq R/m_1$.

11. To see how Eq. (4) approximates the original transformation near a point x_0, y_0 at which $J \neq 0$, let the image of a point on a circle $\Delta x^2 + \Delta y^2 = r^2$ be u, v under $u = u(x,y)$, $v = v(x,y)$, and let its image under Eq. (4) on the corresponding ellipse of Fig. 12 be u_1, v_1. Then it follows from Eq. (3) that

$$|u - u_1| = |\epsilon_1 \Delta x + \epsilon_2 \Delta y| \leq |\epsilon_1 + \epsilon_2|r,$$
$$|v - v_1| = |\epsilon_3 \Delta x + \epsilon_4 \Delta y| \leq |\epsilon_3 + \epsilon_4|r.$$

By identifying $u_x(x_0, y_0)$ in Eq. (4) with the a of Problem 8, and similarly for the other coefficients, we may determine the $m_1 > 0$ of Problem 10, such that $r \leq R/m_1$, where here $R = \sqrt{\Delta u^2 + \Delta v^2}$. Deduce that

$$|u - u_1| \leq \frac{\eta}{\sqrt{2}} \sqrt{\Delta u^2 + \Delta v^2},$$

$$|v - v_1| \leq \frac{\eta}{\sqrt{2}} \sqrt{\Delta u^2 + \Delta v^2}$$

and the distance

$$\sqrt{(u - u_1)^2 + (v - v_1)^2} \leq \eta \sqrt{\Delta u^2 + \Delta v^2}$$

for any given positive fraction η, in the neighborhood of x_0, y_0 such that each of the four ϵ's is less than $m_1 \eta / 2\sqrt{2}$.

12. Let $x = a'u + b'v$, $y = c'u + d'v$ be the equations which solve $u = ax + by$, $v = cx + dy$. Then by Problem 8, $a' = d/D$, $b' = -b/D$, $c' = -c/D$, $d' = a/D$, where $D = ad - bc \neq 0$. Verify that $1 = a'a + b'c$, $0 = a'b + b'd$, $0 = c'a + d'c$, $1 = c'b + d'd$. Show that these equations uniquely determine a', b', c', d' since $D \neq 0$.

13. By identifying u_x with a, x_u with a', etc. in Problem 12 and using Eq. (8), show that the linear approximation for the inverse transformation at u_0, v_0 is the system of simultaneous equations which solve the approximating linear equations (4).

14. From Problems 12 and 13 show that if $u = u(x,y)$, $v = v(x,y)$ and $x = x(u,v)$, $y = y(u,v)$ are inverse transformations, the Jacobian of x and y with respect to u and v is the reciprocal of that of u and v with respect to x and y,

$$\begin{vmatrix} x_u & x_v \\ y_u & y_v \end{vmatrix} \begin{vmatrix} u_x & u_y \\ v_x & v_y \end{vmatrix} = 1,$$

or $JJ' = 1$.

15. From Problem 14 deduce that the solving transformation of the implicit function theorem has a Jacobian $J' = 1/J \neq 0$ at u_0, v_0. Hence this solving transformation $x = x(u,v)$, $y = y(u,v)$ is one-one in some neighborhood of u_0, v_0.

32. Conformal transformations

We now require that the function $f(z)$ of Eq. (1) be analytic in some domain D, as in Section 14. Thus for any point z_0 interior to D the derivative $f'(z_0)$ will exist and we may write

$$f'(z_0) = \lim_{z \to z_0} \frac{w - w_0}{z - z_0}. \tag{7}$$

As stated in Sections 16 and 17, the derivatives u_x, u_y, v_x, v_y must all be continuous in D, and satisfy the Cauchy-Riemann equations

$$u_x = v_y, \quad u_y = -v_x. \tag{8}$$

Hence the value of the Jacobian of Eq. (5) is

$$J = u_x v_y - u_y v_x = u_x^2 + u_y^2 = |u_x + iv_x|^2 = |f'(z)|^2. \tag{9}$$

Suppose that $f'(z_0) \neq 0$. Then $J \neq 0$ at x_0, y_0 and by the theorem quoted in Section 31, there is a neighborhood $|z - z_0| < h$ in which the transformation $w = f(z)$ is one-one. We shall refer to this neighborhood as the domain D_0.

If the inverse function $F(w) = x(u,v) + iy(u,v)$, then

$$F'(w_0) = \frac{1}{f'(z_0)}, \tag{10}$$

by the theorem of Section 25, with roles of z and w interchanged.

Now let C be a smooth curved arc in D_0 which starts at z_0, Fig. 13,

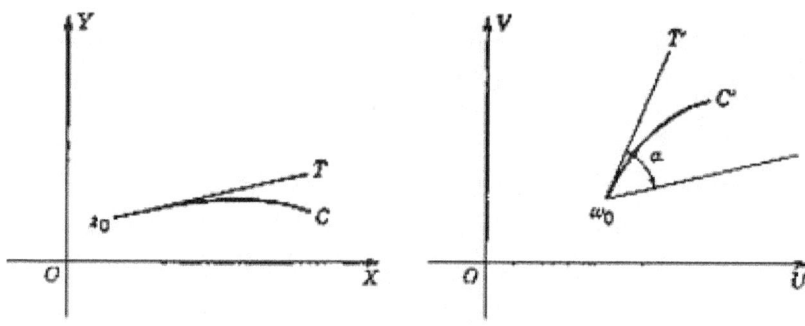

Fig. 13

and $z_0 T$ its tangent at z_0. Then we can show that C', the image of C, also has a tangent $w_0 T'$ at $w_0 = f(z_0)$, and the direction of $w_0 T'$ is that of $z_0 T$ rotated through the angle $\alpha = \arg f'(z_0)$.

To prove this, we first introduce the polar form

$$f'(z_0) = ae^{i\alpha}. \tag{11}$$

Here the value of $\arg f'(z_0) = \alpha$ is determined to within a multiple of 2π, since $f'(z_0) \neq 0$. We next observe that for any choice of the arguments on the right,

$$\arg (w - w_0) = \arg (z - z_0) + \arg \frac{w - w_0}{z - z_0} \qquad (12)$$

gives a possible value of $\arg (w - w_0)$.

For z on C, the vector $z - z_0$ extends along a chord or secant of C. Hence for $z \to z_0$, its direction approaches that of $z_0 T$. It follows from Eqs. (12), (7), and (11) that

$$\lim_{z \to z_0} \arg (w - w_0) = \arg z_0 T + \alpha. \qquad (13)$$

But for any sequence of points w on C', $\neq w_0$ but tending to w_0, the unique continuous inverse function $F(w)$ gives a sequence of points $z \neq z_0$ but tending to z_0. This shows that

$$\lim_{w \to w_0} \arg (w - w_0) = \arg z_0 T + \alpha. \qquad (14)$$

Now $w - w_0$ extends along a chord or secant of C'. Since Eq. (14) shows that its direction approaches a limit when $w \to w_0$, it follows that C' has a tangent line $w_0 T'$ at w_0, and the relation

$$\arg w_0 T' = \arg z_0 T + \alpha \qquad (15)$$

expresses the rotation of direction stated in the theorem.

Consider next any two smooth curved arcs, C_1 and C_2, each of which starts at z_0. For their tangent lines let $\arg z_0 T_1 = \phi_1$, $\arg z_0 T_2 = \phi_2$. Let C_1' be the image of C_1 and let C_2' be the image of C_2. Then these image arcs each have tangents. Moreover,

$$\phi_1' = \arg w_0 T_1' = \phi_1 + \alpha, \quad \phi_2' = \arg w_0 T_2' = \phi_2 + \alpha. \qquad (16)$$

It follows from these relations that

$$\phi_2' - \phi_1' = \phi_2 - \phi_1. \qquad (17)$$

That is, the angle from C_1' to C_2' is the same as the angle from C_1 to C_2.

A mapping or transformation which preserves the angle between any two smooth arcs through a point, in both magnitude and sense, is said to be conformal at the point. Thus we have proved that:

The mapping determined by $w = f(z)$ is conformal at every point where $f(z)$ is analytic, and the derivative $f'(z) \neq 0$.

It also follows from Eqs. (7) and (11) that

$$\lim_{z \to z_0} \frac{|w - w_0|}{|z - z_0|} = |f'(z_0)| = a. \tag{18}$$

As in Eqs. (13) and (14), let w take a sequence of values on C' with $w \to w_0$ and let z be the corresponding sequence of values on C with $z \to z_0$. Then from the relation of arc to chord for the smooth curve C,

$$\lim_{z \to z_0} \frac{\text{arc } z_0 z}{|z - z_0|} = 1. \tag{19}$$

From the similar relation for the smooth image curve C',

$$\lim_{z \to z_0} \frac{\text{arc } w_0 w}{|w - w_0|} = \lim_{w \to w_0} \frac{\text{arc } w_0 w}{|w - w_0|} = 1. \tag{20}$$

It follows from Eqs. (18) to (20) that

$$\lim_{z \to z_0} \frac{\text{arc } w_0 w}{\text{arc } z_0 z} = \frac{ds'}{ds} = a. \tag{21}$$

This shows that all differential lengths issuing from z_0 are multiplied by the same magnification factor $a = |f'(z_0)|$ by the mapping. We call $a = |f'(z_0)|$ the *magnification* at the point. By Eq. (9), $|f'(z)|^2 = u_x^2 + v_y^2$ is a continuous real function. It is positive at z_0 since $f'(z_0) \neq 0$. Hence there will be a neighborhood of z_0, $|z - z_0| < h_1$, in which it remains positive or $|f'(z)|^2 > 0$, which implies that $f'(z) \neq 0$. Let us take the h which defined D_0 to be less than h_1. Then since $f'(z) \neq 0$ throughout D_0, we may take any point of D_0 in place of z_0. Thus our transformation will be conformal and will preserve scale at every point of D_0. Consider any triangle in D_0 with vertices at z_1, z_2, z_3. The images of the three sides will in general be curved arcs joining the image points w_1, w_2, w_3. This curvilinear triangle will have its three angles equal, respectively, to those of the triangle z_1, z_2, z_3.

In general the coefficient of magnification a and the angle of rotation of tangents α will vary from point to point with $f'(z)$. However, they will change only slightly for a small change in z. Hence if z_1, z_2, z_3 are sufficiently close together the curvilinear triangle w_1, w_2, w_3 will have the ratio of the lengths of its curved sides approximately equal to the ratio of the straight sides of the triangle z_1, z_2, z_3. In fact, the curvilinear triangle will approximate a triangle with straight sides similar to the triangle z_1, z_2, z_3. Hence we may conclude that:

The mapping effected by an analytic function is one-one and conformal in some neighborhood of every point at which the derivative of the function is not zero. The mapping is nearly similar for sufficiently small figures inside this neighborhood.

We shall show in Section 68 that the function $w = f(z)$ must map the interior of a region R bounded by a simple closed contour C in the z-plane in a one-one manner onto the interior of a region R' bounded by a simple closed contour C' in the w-plane provided that $w = f(z)$ is analytic in R and on C, and maps C traversed once positively on C' traversed once positively. Although such a mapping is nearly similar for small figures, the shape of a large figure may be very much distorted.

Let $f'(z)$ be zero at z_0, but distinct from zero in some deleted neighborhood of z_0. That is,

$$f'(z_0) = 0, \quad f'(z) \neq 0 \quad \text{if } 0 < |z - z_0| < h. \tag{22}$$

When this relation holds we call z_0 an *isolated* zero of $f'(z)$. As shown in Problem 5 of Exercise 17, under the transformation $w = f(z)$, angles at z_0 are multiplied by a positive integer greater than unity. Since the conformal character of $w = f(z)$ breaks down at z_0, the point z_0 is a *critical point* of the transformation.

We sometimes apply the terms *conformal mapping* and *conformal transformation* to the transformation of points z to points w by means of any nonconstant analytic function $w = f(z)$, even when our discussion includes certain critical points.

A mapping or transformation which preserves angles at a point in magnitude, with or without reversal of sense, is said to be *isogonal*. Since the transformation

$$w = \bar{z} = x - iy \tag{23}$$

is a reflection in the real axis, under it every angle is transformed into one of equal magnitude but opposite sign. If this is followed by a conformal transformation, $w = f(z)$, the resulting transformation

$$w = f(\bar{z}) \tag{24}$$

will be isogonal with reversal of sense at z_0 if $f'(\bar{z}_0) \neq 0$.

88   CONFORMAL TRANSFORMATIONS   Chap. 5

We have seen that in general analytic functions lead to conformal maps. Conversely we may show that under certain restrictions every continuously differentiable conformal mapping of a two-dimensional region is associated with an analytic function.

To establish this converse result, we start with a one-one mapping of some domain D of the z-plane onto a domain D' of the w-plane, determined by two real functions $u(x,y)$ and $v(x,y)$. We assume that the mapping is conformal at each point of D. We assume also that at each point x,y of D the functions u and v each have continuous first partial derivatives and a nonvanishing Jacobian, so that $J = u_x v_y - u_y v_x \neq 0$.

The equations between differentials at x_0, y_0,

$$du = u_x(x_0,y_0)\, dx + u_y(x_0,y_0)\, dy,$$
$$dv = v_x(x_0,y_0)\, dx + v_y(x_0,y_0)\, dy, \qquad (25)$$

define a one-one correspondence of z-slopes, or ratios of dy to dx, and w-slopes, or ratios of dv to du, since $J \neq 0$ at x_0, y_0. This is the same as the transformation of directions by Eq. (4). From the relation of Eq. (4) to Eq. (3), under $u = u(x,y)$, $v = v(x,y)$ a curve through x_0, y_0 with slope dy/dx (which may be infinite, $dy:dx = 1:0$) will have as its image a curve through u_0, v_0 with slope dv/du. Hence, since the given mapping is conformal, so is the linear mapping of Eq. (4) or (25).

A nondegenerate linear mapping of Eq. (4) takes a coordinate grid of equal squares into one based on equal parallelograms or rectangles. Now assume that angles are preserved. Then for the image parallelogram of a square, each vertex angle, as well as the angle of its diagonals, must be a right angle. Hence the image is a square, and square grids go into square grids. Thus the mapping of Eq. (4), which takes z_0 into w_0, must preserve shapes with preservation of sense in order to be conformal. Hence, as a transformation from $z - z_0$ to $w - w_0$, Eq. (4) must be equivalent to a uniform magnification and a positive rotation and so have the form

$$u - u_0 = a[(x - x_0)\cos \alpha - (y - y_0)\sin \alpha],$$
$$v - v_0 = a[(x - x_0)\sin \alpha + (y - y_0)\cos \alpha]. \qquad (26)$$

A comparison of Eqs. (4) or (25) with Eq. (26) shows that

$$u_x = a \cos \alpha = v_y, \quad u_y = -a \sin \alpha = -v_x. \qquad (27)$$

Thus the Cauchy-Riemann equations (8) are satisfied. Since this holds throughout the two-dimensional open region D in which the partial derivatives are assumed to be continuous, by Theorem 4 of Section 16 the function $f(z) = u(x,y) + iv(x,y)$ is analytic in D, as was to be proved.

Exercise 17

1. That the families of curves $u(x,y) = c_1$ and $v(x,y) = c_2$ are *orthogonal*, or intersect at right angles, at any point of intersection where $f'(z) \neq 0$ was proved in Problem 11 of Exercise 8. Give a new proof of this based on the conformal property.

2. From Eqs. (8) and (6) deduce that

$$x_u = \frac{u_x}{J}, \quad x_v = \frac{v_x}{J}, \quad y_u = -\frac{v_x}{J}, \quad y_v = \frac{u_x}{J},$$

where $J = u_x^2 + v_x^2$. Use this to give a new proof that the inverse function $z = x(u,v) + iy(u,v)$ satisfies the Cauchy-Riemann relations $x_u = y_v, x_v = -y_u$, and that $x_u + iy_u = 1/(u_x + iv_x)$, which is equivalent to Eq. (10).

3. Let Arg denote the principal value, $-\pi < \text{Arg}(z - z_0) \leq \pi$. And in Eqs. (12) and (13) let $\lim_{z \to z_0} \text{Arg}(z - z_0) = \phi$. Show that $\phi = \text{Arg } z_0 T$ if $\phi \neq -\pi$, but that $\phi = \text{Arg } z_0 T - 2\pi$ if $\phi = -\pi$.

Let $f(z)$ be analytic at z_0. Then, as will be shown in Section 53, for some $h > 0$, and all z such that $|z - z_0| < h$, there is a power series expansion

$$f(z) = \sum_{n=0}^{\infty} a_n(z - z_0)^n.$$

Anticipating this result, prove that

4. If $f'(z_0) = 0$, and $f(z)$ is not a constant for $|z - z_0| < h$, then z_0 is an isolated zero of $f'(z)$. *Hint:* By Section 21,

$$f'(z) = \sum_{n=1}^{\infty} na_n(z - z_0)^{n-1}.$$

And by Problem 12 of Exercise 10, if $f'(z)$ has zeros at a sequence $z_n \to z_0$, $f'(z) \equiv 0$. But by Theorem 2 of Section 16, this would make $f(z) \equiv a_0$ for $|z - z_0| < h$, contrary to the hypothesis that $f(z)$ was not a constant.

5. If $f'(z_0) = 0$, and z_0 is an isolated zero, for some $n > 1$,

$$f(z) = a_0 + a_n(z - z_0)^n + \ldots, \text{ with } a_n = ae^{i\alpha} \neq 0.$$

Deduce that $w - w_0 = a_n(z - z_0)^n + \ldots$. And as in Eqs. (13) and (14),

$$\lim \arg (w - w_0) = \alpha + n \lim \arg (z - z_0), \quad \phi_1' = \alpha + n\phi_1,$$

so that $\phi_2' - \phi_1' = n(\phi_2 - \phi_1)$.

6. Show that if $f(z)$ is analytic and $f'(z_0) \neq 0$, the transformation $w = \overline{f(z)}$ will be isogonal with reversion of sense at z_0.

7. Let $w = g(z)$ be any transformation which is isogonal with reversal of sense. Show that there are conformal transformations $w = f_1(z)$ and $w = f_2(z)$ such that $g(z) = f_1(\bar{z}) = \overline{f_2(z)}$. *Hint:* Note that the w to $\bar{z}$ and the $\bar{w}$ to z maps are each conformal.

8. Check the form of Eq. (26) by setting $r' = ar$, $\theta' = \theta + \alpha$ in $u - u_0 + i(v - v_0) = r' \operatorname{cis} \theta'$ where $x - x_0 + i(y - y_0) = r \operatorname{cis} \theta$.

9. Proceeding as in Problem 8, but with $r' = ar$, $\theta' = -\theta - \alpha$, derive

$$u - u_0 = a[(x - x_0)\cos \alpha - (y - y_0)\sin \alpha],$$
$$v - v_0 = -a[(x - x_0)\sin \alpha + (y - y_0)\cos \alpha],$$

as one form of Eq. (4) for a similarity transformation with reversal of sense.

10. At x_0, y_0 let u_x, u_y, v_x, v_y be continuous and $u_x v_y - u_y v_x = J \neq 0$. But let the map for $u = u(x,y)$, $v = v(x,y)$ be isogonal with *reversal* of sense. Then the argument which led to Eq. (26) now gives the equations of Problem 9. Deduce that $u_x = -v_y$, $u_y = v_x$, and that $J = -u_x^2 - u_y^2 = -a^2 < 0$.

11. In the assumptions preceding Eq. (25), let us replace the word *conformal* by *isogonal*. Since $J \neq 0$ in D, and it is continuous, it is of fixed sign. By Eq. (27) and Problem 10, for $J > 0$ in D the transformation is conformal and $w = f(z)$ with f analytic as in the text, but for $J < 0$ in D the transformation is isogonal with reversal of sense in D, and for some analytic function f_2 as in Problem 7, $w = \overline{f_2(z)}$.

12. Since the argument which proved that Eq. (4) or (25) represents a shape-preserving transformation merely required that perpendicular directions at z_0 go into perpendicular directions at w_0, verify that we may use this hypothesis for each point in place of *isogonality* in drawing the conclusions of Problem 11.

13. The transformation $u = x^3 + 4xy^2$, $v = x^2y + 4y^3$ was shown to be one-one in Problem 4 of Exercise 16. Show that this transformation is conformal at $z_0 = 0$, and that $f'(0) = 0$. This does not contradict Problem 5, since $f'(z)$ does *not* exist if $z \neq 0$, so that $f(z)$ is not analytic at $z_0 = 0$.

14. Reversing the reasoning of the text, from $f'(z_0) = ae^{i\alpha}$, derive Eq. (27) and then Eq. (26) as the equivalent of Eq. (4). Thus obtain a new proof of the fact that if $f'(z_0) \neq 0$, the mapping by the analytic function $w = f(z)$ is conformal at z_0 with rotation angle α, and magnification factor a independent of direction at z_0.

33. Boundary value problems

We showed in Section 17 that each analytic function of a complex variable $f(z) = u + iv$ determines two real harmonic functions $u(x,y)$ and $v(x,y)$. That is, if $f(z)$ is analytic for z in a domain D, each of the functions u and v has continuous partial derivatives of the second order and is a solution of Laplace's equation

$$\frac{\partial^2 \phi}{\partial x^2} + \frac{\partial^2 \phi}{\partial y^2} = 0 \qquad (28)$$

As indicated in Problems 1, 6, and 23 of Exercise 8, in a suitably restricted region each harmonic function can be associated with an analytic function which determines it as the function u or v.

In many physical applications we encounter the problem of finding a function of two real variables which is harmonic in a given region and satisfies given conditions on the boundary of the region. In some simple cases, we may find the desired function directly from some analytic function with which we are sufficiently familiar, but in other cases, the problem may be simplified by a conformal mapping which transforms the given region into a simpler one. This is effective because under a conformal transformation from the z-plane to the w-plane, Laplace's equation in x and y corresponds to Laplace's equation in u and v.

To describe this in detail, let $U(x,y)$ denote any harmonic function of x and y in a domain D of the z-plane. We assume that D is simply connected, with boundary a simple closed curve C. Let $w = u + iv = f(z)$ be an analytic function for z in D, with $f'(z) \neq 0$, which maps D and C on a simply connected domain D' with boundary C' of the w-plane in a one-one manner. Then $z = x + iy = f^{-1}(w)$, the inverse analytic function which maps D' on D, determines two real conjugate harmonic functions $x(u,v)$ and $y(u,v)$. We may use these to find $U_1(u,v) = U[x(u,v), y(u,v)]$. We wish to show that $U_1(u,v)$, which corresponds to $U(x,y)$ under the conformal mapping, is harmonic in D'.

We first observe that since D is simply connected, and $U(x,y)$ is harmonic in D, we may find a conjugate function $V(x,y)$ such that $U + iV = F(z)$ is an analytic function of z in D. Then since $f^{-1}(w)$ is analytic for w in D', and $z = f^{-1}(w)$ maps D' on D, it follows that the composite function $F[f^{-1}(w)] = U_1 + iV_1$ is analytic in D'. Hence its real part,

$$U_1(u,v) = U[x(u,v), y(u,v)], \qquad (29)$$

is harmonic in D', as was to be proved. Briefly, our result is

Under any one-one conformal transformation corresponding to an analytic function with nonzero derivative, a harmonic function in either plane is transformed into a harmonic function in the other plane.

Any given boundary conditions on C will determine corresponding boundary conditions on C'. Let u_0, v_0 on C' be the image of x_0, y_0 on C. Then $U_1(u_0, v_0) = U(x_0, y_0)$, which determines the values on C' if the values on C are given. Finding a function which is harmonic in D and takes prescribed values on C is known as *Dirichlet's problem*. Thus under the conformal transformation, Dirichlet's problem for $U(x,y)$ in D transforms into Dirichlet's problem for $U(u,v)$ in D'.

If the boundary value is not prescribed at an isolated point of discontinuity for the boundary values, or at infinity, the condition that the harmonic function be uniformly bounded in some neighborhood of this point makes Dirichlet's problem determinate.

Instead of prescribing the value of U, we may require that $dU/dn = 0$ on an arc of C, where dU/dn is the inner normal derivative of U. Before transforming such conditions, we consider the transformation of a directional derivative dU/ds under a conformal transformation. We recall that the directional derivative at z_0 in the direction ϕ is the ordinary derivative with respect to arc length along the straight line $z = z_0 + s\operatorname{cis}\phi$, or along any curve through z_0 with positive direction tangent to this line. Let ϕ' be the direction at w_0 corresponding to ϕ and dU_1/ds' be the corresponding directional derivative of U_1. Then since $U_1 = U$,

$$\frac{dU_1}{ds'} = \frac{dU}{ds'} = \frac{dU}{ds}\frac{ds}{ds'}. \tag{30}$$

From Eqs. (21) and (18), we have

$$\frac{ds'}{ds} = a = |f'(z_0)| = \left|\frac{dw}{dz}\right|. \tag{31}$$

Hence for s and s' along corresponding directions, we have

$$\frac{dU_1}{ds'} = \frac{dU/ds}{|dw/dz|}. \tag{32}$$

Now suppose that our transformation is conformal at z_0 on C with image w_0 on C'. Then the direction of the inward normal at C maps onto that of the inward normal at C'. Hence, for the normal derivatives, we have as in Eq. (32):

$$\frac{dU_1}{dn'} = \frac{dU/dn}{|dw/dz|}. \tag{33}$$

It follows from this that a boundary condition $dU/dn = 0$ goes into one of similar type.

The more general condition $dU/dn + hU = g$, with $h > 0, g > 0$ becomes

$$\frac{dU_1}{dn'} + h\left|\frac{dw}{dz}\right|U_1 = g\left|\frac{dw}{dz}\right|. \tag{34}$$

Since dw/dz will in general vary along C', an original condition with h and g constant becomes a similar condition with these constants replaced by variable functions of u,v on C', which is an added complication.

Exercise 18

1. Let $u(x,y)$, $v(x,y)$, and $U_1(u,v)$ be *any* functions with continuous second partial derivatives. Verify that if

$$U(x,y) = U_1[u(x,y),v(x,y)],$$

$$\frac{\partial^2 U}{\partial x^2} = \frac{\partial^2 U_1}{\partial u^2}u_x^2 + 2\frac{\partial^2 U_1}{\partial u\,\partial v}u_x v_x + \frac{\partial^2 U_1}{\partial v^2}v_x^2 + \frac{\partial U_1}{\partial u}u_{xx} + \frac{\partial U_1}{\partial v}v_{xx}.$$

2. From Problem 1, and the similar expression with x replaced by y throughout, deduce that if $f(z) = u + iv$ is analytic,

$$\frac{\partial^2 U}{\partial x^2} + \frac{\partial^2 U}{\partial y^2} = \left(\frac{\partial^2 U_1}{\partial u^2} + \frac{\partial^2 U_1}{\partial v^2}\right)(u_x^2 + v_x^2)$$
$$= \left(\frac{\partial^2 U_1}{\partial u^2} + \frac{\partial^2 U_1}{\partial v^2}\right)\left|\frac{dw}{dz}\right|^2.$$

3. From Problem 2, deduce an alternative proof of the statement following Eq. (29).

4. Verify that if $U_1(u,v) = u^2 + v^2$,

$$\frac{\partial^2 U_1}{\partial u^2} + \frac{\partial^2 U_1}{\partial v^2} = 4.$$

From this and Problem 2 deduce that if $f(z)$ is analytic,

$$\left(\frac{\partial^2}{\partial x^2} + \frac{\partial^2}{\partial y^2}\right)|f(z)|^2 = 4|f'(z)|^2.$$

5. From the analytic character of

$$\log w = \mathrm{Log}\sqrt{u^2 + v^2} + i\tan^{-1}\frac{v}{u},$$

deduce that $\mathrm{Log}|f(z)|$ is harmonic in any domain D in which $f(z)$ is analytic and $f(z) \neq 0$, while $\arg f(z)$ is also harmonic in this D if D is also simply connected. Anticipating the result of Section 53 that $f'(z)$ is analytic whenever $f(z)$ is, deduce that $\mathrm{Log}|f'(z)|$ is harmonic in any domain in which $f(z)$ is analytic and $f'(z) \neq 0$.

6. Verify that if $U_1(u,v) = -\mathrm{Log}(1 - u^2 - v^2)$, then

$$\frac{\partial^2 U_1}{\partial u^2} + \frac{\partial^2 U_1}{\partial v^2} = \frac{4}{(1 - u^2 - v^2)^2}.$$

Deduce from this, the last result of Problem 5, and Problem 2 that if

$$g(x,y) = \mathrm{Log}\frac{|f'(z)|}{1 - |f(z)|^2}, \quad \text{then } \frac{\partial^2 g}{\partial x^2} + \frac{\partial^2 g}{\partial y^2} = 4e^{2g}$$

in any domain where $f(z)$ is analytic, $f'(z) \neq 0$, and $|f(z)| \neq 1$. *Hint:* Write $g(x,y) = \mathrm{Log}|f'(z)| - \mathrm{Log}(1 - u^2 - v^2)$.

The analytic character of $\log z = \mathrm{Log}\, r + i\theta$ shows that for $z \neq 0$, $\mathrm{Log}\, r = \mathrm{Log}\sqrt{x^2 + y^2}$ and $\theta = \tan^{-1}(y/x)$ are harmonic functions. Deduce that:

7. The function H which is harmonic for $a < r < b$, has $H = A$ for $r = a$ and $H = B$ for $r = b$ is

$$H = \frac{(B - A) \log r + A \log b - B \log A}{\log b - \log a}.$$

8. The function H which is harmonic and uniformly bounded in the sector $0 < \theta < \alpha$, $0 < r < b$, has $H = 0$ for $\theta = 0$, $H = A$ for $\theta = \alpha$, and $dH/dn = -dH/dr = 0$ for $r = b$ is $H = A\theta/\alpha$.

For a real,

$$-\arg \frac{a - z}{a + z} = \tan^{-1} \frac{2ay}{a^2 - x^2 - y^2}$$

is harmonic in any simply connected domain D which does not include either of the two points $z = \pm a$ in its interior, by Problem 5. By checking the boundary conditions in each case, verify that:

9. The function H which is harmonic and uniformly bounded inside the circle $x^2 + y^2 = a^2$ has $H = A$ on the upper semicircle with $y < 0$, and has $H = B$ on the lower semicircle with $y < 0$ is

$$H = \frac{A + B}{2} + \frac{A - B}{\pi} \tan^{-1} \frac{2ay}{a^2 - x^2 - y^2},$$

where the value of the inverse tangent is taken in the interval $-\pi/2 \leq \theta_1 \leq \pi/2$.

10. The function H which is harmonic and uniformly bounded in the upper half plane with $y > 0$, has $H = A$ for $y = 0$, $|x| < a$, and has $H = B$ for $y = 0$, $|x| > a$, is

$$H = A + \frac{B - A}{\pi} \tan^{-1} \frac{2ay}{a^2 - x^2 - y^2},$$

where the inverse tangent is taken as $0 \leq \theta_2 \leq \pi$ to make θ_2 continuous in the upper half plane.

11. The function H which is harmonic and uniformly bounded inside the semicircular domain $x^2 + y^2 < a^2$, $y > 0$, has $H = A$ on the diameter $y = 0$, and has $H = B$ on the semicircle is

$$H = A + \frac{2(B - A)}{\pi} \tan^{-1} \frac{2ay}{a^2 - x^2 - y^2}.$$

Here for the positive quantity the inverse tangent may be taken as $0 \leq \theta_3 \leq \pi/2$.

12. By combining Problem 11 applied to the u,v plane with the transformation $w = z^\omega = r^\omega \cos \omega\theta + i r^\omega \sin \omega\theta$, verify that the function H which is harmonic and uniformly bounded in the sector $0 < \theta < \alpha$, $0 < r < a$, which has $H = A$ on each bounding radius, $\theta = 0$ and $\theta = \alpha$, and which has $H = B$ on the arc $r = a$, is

$$H = A + \frac{2(B - A)}{\pi} \tan^{-1} \frac{2ar^\omega \sin \omega\theta}{a^2 - r^{2\omega}}$$

with $\omega = \pi/\alpha$. Here $0 \leq \theta_1 \leq \pi/2$.

13. If $w = \sin z = \sin x \cosh y + i \cos x \sinh y$, $w'(z) = \cos z \neq 0$ for $|z| < \pi/2$. Deduce from this that for $0 < x < \pi/2$, $\arg \sin z = \tan^{-1} \frac{\tanh y}{\tan x}$ is harmonic.

14. Using Problem 13 with z replaced by $\pi z/2a$, and also with z replaced by $\pi/2 - \pi z/2a$, to prove that the function is harmonic, verify that the function H which is harmonic and uniformly bounded in the strip $0 < x < a$, $y > 0$ with $H = A$ for $x = a$, $H = B$ for $y = 0$, and $H = C$ for $x = 0$ is

$$H = B + \frac{2(C - B)}{\pi} \tan^{-1} \frac{\tanh \frac{\pi y}{2a}}{\tan \frac{\pi x}{2a}} + \frac{2(A - B)}{\pi} \tan^{-1} \frac{\tanh \frac{\pi y}{2a}}{\cot \frac{\pi x}{2a}},$$

with each inverse tangent in the first quadrant.

15. Verify that if $C = A$ in Problem 14, the solution reduces to

$$H = B + \frac{2(A - B)}{\pi} \tan^{-1} \frac{\sinh \frac{\pi y}{a}}{\sin \frac{\pi x}{a}}.$$

16. The only harmonic function H which is uniformly bounded for $x > \pi/2$ and which $= 0$ for $x = \pi/2$ is $H = 0$. From the analytic character of $e^{ix} = e^{-y} \cos x + i e^{-y} \sin x$, deduce that $H = A e^{-y} \cos x$ is harmonic in any finite domain. Verify that for any $A \neq 0$, $|H| \to \infty$ for $x = 2\pi$, $y \to -\infty$, so that $H = A e^{-y} \cos x$ is an unbounded "harmonic" function for $x > \pi/2$, with $H = 0$ for $x = \pi/2$.

17. From Problem 16 applied to the u,v plane, $e^{-v} \cos u$ is harmonic in any finite domain. Combining this with the transformation

$$w = \frac{1}{z} = \frac{x}{x^2 + y^2} - \frac{iy}{x^2 + y^2},$$

deduce that the function

$$H = \exp\frac{y}{x^2+y^2} \cos\frac{x}{x^2+y^2}$$

is harmonic for all x,y except $0,0$. Verify that for any point except this on the circle $x^2 + y^2 - 2x/\pi = 0$, $H = 0$. Note that for $x = 0$, $y \to 0+$, $H \to \infty$. Hence for any complex $A \neq 0$,

$$H = A \exp\frac{y}{x^2+y^2} \cos\frac{x}{x^2+y^2}$$

is an unbounded "harmonic" function inside the circle $x^2 + y^2 - 2x/\pi = 0$ which takes on the value zero at each point on the circle except $0,0$.

6

BILINEAR TRANSFORMATIONS

Let w_0 be the image of z_0 under a conformal transformation. Then we have seen that for suitable sufficiently small neighborhoods of these points, the transformation is one-one and approximately shape-preserving. We first show that for the *linear* transformation $w = Az + B$, with $A \neq 0$, the transformation is one-one and exactly shape-preserving for the entire finite plane. We then make a suitable convention about infinity, and define the closed plane as consisting of the finite plane together with the point at infinity. This leads us to discuss those conformal transformations which are one-one throughout the closed plane. These are the *bilinear* transformations,

$$w = \frac{Az + B}{Cz + D}, \quad \text{with } AD - BC \neq 0.$$

Although they do not preserve large shapes, they do take "circles," including straight lines as special cases, into "circles." The geometric properties of bilinear transformations are described in terms of geometrical inversion and families of coaxial circles. Finally we show how to find those particular bilinear transformations which take a specified figure in the z-plane into a specified figure of suitable type in the w-plane.

34. Linear transformations

The general *linear* transformation is defined by a relation of the form

$$w = Az + B, \qquad A \neq 0 \tag{1}$$

where A and B are complex constants. Our requirement on A implies that

$$\frac{dw}{dz} = A \neq 0, \tag{2}$$

so that the transformation is conformal.

We begin with the special case with $A = 1$,

$$w = z + B. \tag{3}$$

Here w, the image of z, may be found from z and B by vectorial addition, as in Section 6. Since each point z undergoes the same vectorial displacement B, the effect of the transformation (3) on any figure of the z-plane is merely a sliding parallel to itself, or a *translation*. The translated image figure has the same size, shape, and orientation as the original figure. For $B = 0$, Eq. (3) is the identity.

We next let $B = 0$ in Eq. (1). Then if $A = a \operatorname{cis} \alpha$ and $z = r \operatorname{cis} \theta$, we may write

$$w = Az = are^{i(\theta + \alpha)} = ar \operatorname{cis}(\theta + \alpha). \tag{4}$$

This shows that a point z with polar coordinates r, θ has an image point w with polar coordinates $ar, \theta + \alpha$. Hence the radius vector is rotated about the origin through $\alpha = \arg A$, and stretched by the magnification factor $a = |A|$. By this *rotation* and *magnification*, each figure is transformed into a figure of the same shape. That the rotation and magnification is the same for all points is consistent with Section 32, since $dw/dz = A$. For $A = 1$, Eq. (4) is the identity.

Finally, consider the general relation of Eq. (1). Then

$$w = Az + B, \quad \text{if} \quad W = Az, \quad w = W + B. \tag{5}$$

Thus the first mapping may be effected by the rotation and magnification which takes z to W, followed by the translation which takes W to w. Thus the general linear relation of Eq. (1) exactly preserves the shapes of all figures.

Exercise 19

Sketch the image in the w-plane of the rectangle bounded by the lines $x = 0, x = 2, y = 0, y = 4$ for each of the following linear transformations.

1. $w = -2iz + 5$.
2. $w = (2 + 2i)z + 4i$.
3. $w = iz + 3 + 2i$.
4. $w = (1 + 3i)z - 2i$.

5. Deduce the equations $x' = x - x_0$, $y' = y - y_0$ for a parallel coordinate system with new origin at x_0, y_0 from the transformation $w = z - x_0 - iy_0$.

6. Deduce the equations $x' = x \cos \alpha + y \sin \alpha$, $y' = -x \sin \alpha + y \cos \alpha$ for a coordinate system rotated through the angle α, from the transformation $w = \text{cis}\,(-\alpha)z$.

7. Use the equations $Z = z + B/A$, $w = AZ$ to interpret $w = Az + B$ as a translation *followed* by a rotation and magnification.

8. Show that every transformation $w = Az + B$, $A \neq 0$ with $A \neq 1$ has a fixed point, $w_0 = z_0 = B/(1 - A)$, and it is a rotation and magnification about this point, $w - w_0 = A(z - z_0)$.

9. Verify that the inverse of the transformation of Problem 8 is $z = (1/A)w - B/A$ and has the same fixed point as the direct transformation. Also show that it can be written $z - z_0 = (1/A)(w - w_0)$.

Verify that the transformation inverse to that of

10. Problem 1 is $z = (i/2)w - 5i/2$.
11. Problem 3 is $z = -iw - 2 + 3i$.

35. Bilinear transformations

The general *bilinear* transformation is defined by a relation of the form

$$w = \frac{Az + B}{Cz + D}, \qquad AD - BC \neq 0, \tag{6}$$

where A, B, C, D are complex constants.

First let $C = 0$. Then $AD - BC = AD \neq 0$, so that $A \neq 0$ and $D \neq 0$. Hence $w = (A/D)z + B/D$ with $A/D \neq 0$, which has the form of Eq. (1).

Now assume that $C \neq 0$. Then by division we may deduce that

$$w = \frac{A}{C} - \frac{AD - BC}{C} \frac{1}{CZ + D}, \quad \frac{dw}{dz} = \frac{AD - BC}{(Cz + D)^2}. \tag{7}$$

Hence, if $z \neq -D/C$, our requirement on $AD - BC$ implies a derivative $dw/dz \neq 0$ and the transformation is conformal.

The first part of Eq. (7) also shows that our transformation (6) is equivalent to

$$w = -\frac{AD - BC}{C} W + \frac{A}{C}, \quad W = \frac{1}{Z}, \quad Z = Cz + D. \tag{8}$$

The first and third of these relations have the form of Eq. (1), and so are transformations which preserve shape. Thus the second transformation, which has the form of $w = 1/z$, will determine how shapes are changed by the original transformation (6) when $C \neq 0$. Accordingly, we shall next consider the transformation $w = 1/z$.

36. The transformation $w = 1/z$

If $z = r \operatorname{cis} \theta$, we have

$$w = \frac{1}{z} = \frac{1}{re^{i\theta}} = \frac{1}{r} e^{-i\theta} = \frac{1}{r} \operatorname{cis}(-\theta). \tag{9}$$

Hence a point z with polar coordinates r, θ has an image point w with polar coordinates $1/r$, $-\theta$. We may conveniently effect this transformation in two stages.

The first stage takes $z = r \operatorname{cis} \theta$ to $W = (1/r) \operatorname{cis} \theta$, the point with the same θ but with reciprocal r. Since the two points lie on the same extended radius of the unit circle, and their distances from the center of the circle have $|z| \cdot |w| = 1$, the square of the radius, this stage is an *inversion* in the unit circle.

The second stage takes

$$W = \frac{1}{r} \operatorname{cis} \theta \quad \text{into} \quad w = \frac{1}{r} \operatorname{cis}(-\theta),$$

the point with the same absolute value but with negative argument. Hence w is the conjugate of W, and the second stage is a reflection in the real axis. We have

$$w = \overline{W}, \quad W = \frac{1}{r} \operatorname{cis} \theta = \frac{1}{r} e^{i\theta} = \frac{1}{\bar{z}}. \tag{10}$$

By Section 32, for $z \neq 0$, each of these transformations is *isogonal*, preserving angles in magnitude but reversing their sense. They combine to give the conformal transformation of Eq. (9).

The inversion takes all points outside the unit circle, with $r > 1$, into points inside the circle, with $r < 1$, and conversely except for the center. Hence it must change large shapes drastically. However we may show that it takes any *"circle,"* that is straight line or circle, into a "circle." To show this, we begin by noting that the equation of a "circle" is

$$a(x^2 + y^2) + 2bx + 2cy + d = 0, \quad b^2 + c^2 > ad. \quad (11)$$

When $a = 0$, the equation represents a straight line since the condition $b^2 + c^2 > ad$ prevents b and c from both vanishing. And when $a \neq 0$, the condition makes

$$R^2 = \left(\frac{b}{a}\right)^2 + \left(\frac{c}{a}\right)^2 - \frac{d}{a} > 0,$$

so that the radius of the circle is real and positive. Since $z = x + iy$, $\bar{z} = x - iy$, Eq. (11) may be rewritten

$$az\bar{z} + b(z + \bar{z}) + ci(\bar{z} - z) + d = 0. \quad (12)$$

With $S = b - ci$, $\bar{S} = b + ci$, this becomes

$$az\bar{z} + Sz + \bar{S}\bar{z} + d = 0, \quad S\bar{S} > ad. \quad (13)$$

The inversion corresponds to the following relations:

$$W = \frac{1}{z}, \quad \bar{W} = \frac{1}{\bar{z}}, \quad z = \frac{1}{W}, \quad \bar{z} = \frac{1}{\bar{W}}. \quad (14)$$

Hence when z satisfies Eq. (13), W satisfies

$$dW\bar{W} + S\bar{W} + \bar{S}W + a = 0, \quad S\bar{S} > da. \quad (15)$$

Since this has the same form as Eq. (13), it follows that each "circle" goes into a "circle" under $W = 1/z$.

Under the reflection $w = \bar{W}$ all shapes are unchanged, so that in particular a "circle" goes into a "circle." Since this has just been shown to hold for $W = 1/\bar{z}$, it follows that the transformation $w = 1/z$ takes "circles" into "circles." By the remark following Eq. (8), this property also holds for the general bilinear transformation of Eq. (6).

37. The point at infinity

If $w = 1/z$, then $z = 1/w$. Thus every point $w \neq 0$ is the image of some point z. The points in a deleted neighborhood of $w = 0$, with $0 < |w| < h$ are the images of the points with $|z| > 1/h$. For h small, $1/h$ is large. Thus the points outside of a large circle in the z-plane are mapped by $w = 1/z$ onto the points inside a small neighborhood of $w = 0$. This makes it natural to close the complex plane by a single point at infinity, I or $z = \infty$, which we consider to have $w = 0$ as its image under $w = 1/z$. Any statement about the behavior of a function $f(z)$ at I, or for $z = \infty$, merely means the behavior of $f(1/Z)$ for $Z = 0$.

To illustrate how the use of the point I may enable us to simplify a geometric statement, we consider the question of when a circle or straight line goes into a circle or straight line under the transformation $w = 1/z$. Putting $W = \bar{v}$ in Eq. (15), or $z = 1/w$ in Eq. (13) leads to

$$dw\bar{w} + \bar{S}w + S\bar{w} + a = 0, \quad \bar{S}S > ad. \tag{16}$$

Reference to Eq. (11) shows that Eq. (13) will represent a straight line if $a = 0$, and will have a locus which passes through the origin if $d = 0$. It follows from Eqs. (16) and (13) that under the transformation $w = 1/z$: (a) circles not through the origin go into circles not through the origin; (b) circles through the origin go into straight lines not through the origin; (c) straight lines through the origin go into straight lines through the origin.

Let us call the origin O' and introduce the point at infinity I' in the w-plane. And consider a straight line as a "circle" through I or I'. Then the statements (a), (b), (c) may be summarized by saying that "circles" in the z-plane through O or I go into "circles" in the w-plane through I' or O', respectively.

If $f(z) = 1/z$, then $f(1/Z) = Z$. Since $w = Z$ is conformal at $Z = 0$, we say that $w = 1/z$ is conformal at $z = 0$, with image point I', and similarly at $z = \infty$, or I with image point $w = 0$, or O'.

Let $w = Az + B$, $A \neq 0$ as in Eq. (1). Then $w = \infty$ for $z = \infty$. Hence I' is the image of I. To study the nature of the transformation at this "point," we must put $w = 1/W$ and $z = 1/Z$. This leads to

$$\frac{1}{W} = \frac{A}{Z} + B, \quad W = \frac{Z}{BZ + A}, \quad \frac{dW}{dZ} = \frac{A}{(BZ + A)^2}. \quad (17)$$

For $Z = 0$, $dW/dZ = 1/A \neq 0$, so that the Z-to-W map is conformal at $Z = 0$, with image point $W = 0$. Hence we say that $w = Az + B$ is conformal at I, with image point I'.

Let us use the phrase *closed plane* to emphasize that the plane is considered to be made up of all the finite points plus the one point at infinity. Then our discussion shows that the linear transformation of Eq. (1), as well as the transformation $w = 1/z$, is a one-one conformal transformation of the closed z-plane onto the closed w-plane. It follows from Eq. (8) that this is also true of the general bilinear transformation of Eq. (6).

Exercise 20

1. Let

$$W = \frac{A_1 z + B_1}{C_1 z + D_1} \quad \text{with } A_1 D_1 - B_1 C_1 \neq 0,$$

$$w = \frac{A_2 W + B_2}{C_2 W + D_2} \quad \text{with } A_2 D_2 - B_2 C_2 \neq 0.$$

Verify that $w = \dfrac{Az + B}{Cz + D}$ with

$$A = A_2 A_1 + B_2 C_1,$$
$$B = A_2 B_1 + B_2 D_1,$$
$$C = C_2 A_1 + D_2 C_1,$$
$$D = C_2 B_1 + D_2 D_1,$$

or in terms of matrices,

$$\begin{Vmatrix} A_2 & B_2 \\ C_2 & D_2 \end{Vmatrix} \begin{Vmatrix} A_1 & B_1 \\ C_1 & D_1 \end{Vmatrix} = \begin{Vmatrix} A & B \\ C & D \end{Vmatrix}.$$

Also

$$(A_2 D_2 - B_2 C_2)(A_1 D_1 - B_1 C_1) = AD - BC,$$

or

$$\begin{vmatrix} A_2 & B_2 \\ C_2 & D_2 \end{vmatrix} \begin{vmatrix} A_1 & B_1 \\ C_1 & D_1 \end{vmatrix} = \begin{vmatrix} A & B \\ C & D \end{vmatrix}.$$

Hence $AD - BC \neq 0$.

Sec. 37   BILINEAR TRANSFORMATIONS   105

2. Show that any transformation
$$w = \frac{Az + B}{Cz + D}, \quad AD - BC \neq 0$$
has an equivalent representation
$$w = \frac{A'z + B'}{C'z + D'} \quad \text{with } A'D' - B'C'$$
equal to any given complex number $\neq 0$. Note that if $E \neq 0$, and $A' = EA$, $B' = EB$, $C' = EC$, $D' = ED$, then $A'D' - B'C' = E^2(AD - BC)$. It follows from the result of this problem that either for a single transformation, or for the combination of transformations as in Problem 1, it is only the nonzero nature of $AD - BC$ which is significant.

3. If Eq. (6) holds, then $Cwz + Dw - Az - B = 0$. This is the bilinear relation which gives the transformation its name. Deduce that
$$z = \frac{-Dw + B}{Cw - A}, \quad \text{with } (-D)(-A) - BC = AD - BC \neq 0,$$
so that the transformation of Eq. (6) has an inverse of the same form.

4. Let T_1 and T_2 denote two bilinear transformations, and T_2T_1 denote the combination of the transformations in order as in Problem 1. Verify that with this law of combination the bilinear transformations form a *group* by noting that: (a) $T_2T_1 = T$ *is again a bilinear transformation*, by Problem 1; (b) *the identity* $T_0 : w = z$ *is a particular bilinear transformation*, for $A = D = 1$, $C = B = 0$; (c) *for each T, there is an inverse transformation* T^{-1}, *such that* $T^{-1}T = TT^{-1} = T_0$, *the identity of* (b), by Problem 3.

5. Let Eq. (6) hold, and let $C \neq 0$. Then $z = -D/C$ has $w = \infty$, or P as its image. Verify directly that the transformation is conformal here, since if $w = 1/W$, then
$$W = \frac{Cz + D}{Az + B}, \text{ and at } z = -\frac{D}{C}, \frac{dW}{dz} = -\frac{C^2}{AD - BC}.$$

6. Consider a straight line in the z-plane which does not pass through the origin O. Let the vector from O normal to this line have length p and argument θ_0. Verify that $r \cos(\theta - \theta_0) = p$. If $n_0 = \text{cis } \theta_0$ is the unit normal, $\bar{n}_0 = \text{cis}(-\theta_0)$ and $\text{Re } z\bar{n}_0 = p$, and if $z_0 = sn_0$, with s real but of either sign, $\text{Re } z\bar{z}_0 = sp$. This shows that $\text{Re } z\bar{z}_0 = b$

is the equation of a straight line whose normal from the origin has the direction of bz_0 for any real $b \neq 0$. A consideration of limits shows that $\operatorname{Re} z\bar{z}_0 = 0$ is the equation of a straight line through the origin perpendicular to z_0.

7. Verify that $|z - z_0| = r_0$ is the equation of a circle with center at z_0 and radius r_0. Also that the circle $|z - z_1| = r_1$ cuts the first circle at right angles if $|z_0 - z_1|^2 = r_0^2 + r_1^2$.

In each of Problems 8-12, use the results of Problems 6 and 7 to interpret the equations. And verify that under the inversion transformation of Eq. (14), $z = 1/\overline{W}$,

8. The circle $|z - s| = r$ has as its image the circle
$$\left| W - \frac{s}{s^2 - r^2} \right| = \frac{r}{|s^2 - r^2|}$$
if s is real and $s \neq \pm r$, but the straight line $U = \operatorname{Re} W = 1/2s$ if $s = r$ or $s = -r$.

9. The straight line $x = \operatorname{Re} z = t$ has as its image the circle
$$\left| W - \frac{1}{2t} \right| = \frac{1}{2|t|}$$
if $t \neq 0$, but the straight line $U = \operatorname{Re} W = 0$, essentially itself, if $t = 0$.

10. The straight line $\operatorname{Re} z\bar{z}_0 = t$ has as its image the circle
$$\left| W - \frac{z_0}{2t} \right| = \left| \frac{z_0}{2t} \right|$$
if $t \neq 0$, but the straight line $\operatorname{Re} W\bar{z}_0 = 0$, essentially itself, if $t = 0$.

11. The circle $|z - z_0| = r$ has as its image the circle
$$\left| W - \frac{z_0}{|z_0|^2 - r^2} \right| = \frac{r}{||z_0|^2 - r^2|}$$
if $|z_0| \neq r$, but the straight line $\operatorname{Re} W\bar{z}_0 = \frac{1}{2}$ if $|z_0| = r$.

12. The circle $|z - z_0|^2 = |z_0|^2 - 1 > 0$ has as its image the circle $|W - z_0|^2 = |z_0|^2 - 1$, essentially itself.

13. By Problem 7, the circle of Problem 12 cuts the unit circle $|z| = 1$ at right angles. Use this fact to obtain the result of Problem 12 without any calculation.

14. From Eqs. (13) and (15), under the inversion $W = 1/\bar{z}$ a "circle" will go into itself if and only if $a = d$, or $S = 0$ and $a = -d$.

Check the first condition for Re $z = 0$ in Problem 9, Re $z\bar{z}_0 = 0$ in Problem 10, $|z - z_0|^2 = |z_0|^2 - 1 > 0$ in Problem 12. And $|z| = 1$ from Problem 8 with $s = 0$, $r = 1$, illustrates the second condition.

15. From Eqs. (13) and (16), deduce that a "circle" will go into itself under the conformal transformation $w = 1/z$ if and only if $a = d$ and S is real, or S is pure imaginary and $a = -d$. Thus the "circle" must be $|z - t|^2 = t^2 - 1 > 0$, Re $z = 0$, $|z - it|^2 = t^2 + 1 \geq 1$, Im $z = 0$.

In each of Problems 16 and 17, use the results of Problems 6 and 7 to interpret the equations. And verify that under the conformal transformation $w = 1/z$,

16. The straight line Re $z\bar{z}_0 = t$ has as its image the circle

$$\left| w - \frac{\bar{z}_0}{2t} \right| = \left| \frac{\bar{z}_0}{2t} \right|$$

if $|z_0| \neq r$, but the straight line Re $zz_0 = \tfrac{1}{2}$ if $|z_0| = r$.

17. The circle $|z - z_0| = r$ has as its image the circle

$$\left| w - \frac{\bar{z}_0}{|z_0|^2 - r^2} \right| = \frac{r}{||z_0|^2 - r^2|}$$

if $z_0 \neq r$, but the straight line Re $zz_0 = \tfrac{1}{2}$ if $|z_0| = r$.

Let X, Y, Z be rectangular coordinates in three-space. Then $X^2 + Y^2 + (Z - \tfrac{1}{2})^2 = \tfrac{1}{4}$ is the equation of a sphere through $N = (0,0,1)$ the north pole and tangent to the XY-plane at the origin, the south pole. Let P be any point on the sphere, and $z = x + iy$ denote the point $(x,y,0)$ in the XY-plane. Then if NPz is a straight line, z is called the *stereographic projection* of P on the XY-plane. Let the polar coordinates of z be r,θ. And let the polar coordinates for the X,Y of P be R,θ_1. Then $\theta_1 = \theta$ and $(1 - Z)/1 = R/r$. Also, from the equation of the sphere, $R^2 + Z^2 - Z = 0$. It follows that

$$r^2 = \frac{R^2}{(1 - Z)^2} = \frac{Z}{1 - Z} \quad \text{and} \quad Z = \frac{r^2}{1 + r^2}.$$

Hence $R/r = 1/(1 + r^2)$. Verify the following.

18. $X = \dfrac{x}{1 + x^2 + y^2}$, $Y = \dfrac{y}{1 + x^2 + y^2}$, $Z = \dfrac{x^2 + y^2}{1 + x^2 + y^2}$.

And for $Z \neq 1$,

$$x = \frac{X}{1 - Z}, \quad y = \frac{Y}{1 - Z}, \quad r^2 = \frac{Z}{1 - Z}.$$

108 BILINEAR TRANSFORMATIONS Chap. 6

19. If x,y is on the "circle" of Eq. (11), then X,Y,Z satisfies the equation $2bx + 2cy + (a - d)Z + d = 0$. This represents a plane whose distance from the center of the sphere $(0,0,\tfrac{1}{2})$ is

$$D = \frac{\tfrac{1}{2}|a + d|}{(a - d)^2 + 4(b^2 + c^2)}.$$

But the condition in Eq. (11), $b^2 + c^2 > ad$, makes $(a - d)^2 + 4(b^2 + c^2) > (a + d)^2$. Hence $D < \tfrac{1}{2}$, and the plane cuts the sphere in a real circle.

20. $dx^2 + dy^2 = \dfrac{dX^2 + dY^2 + dZ^2}{(1 - Z)^2}$ for $Z \neq 1$. And

$$dX^2 + dY^2 + dZ^2 = \frac{dx^2 + dy^2}{(1 + x^2 + y^2)^2}.$$

Under the mapping from the plane onto the sphere, the magnification of lengths $\dfrac{dS}{ds} = \dfrac{1}{1 + x^2 + y^2}$ is independent of direction at each point z. Hence this mapping is conformal for all finite values of x and y.

21. When $1/z \to 0$, or $x^2 + y^2 \to \infty$, on the sphere $P \to N$. Thus the point I or $z = \infty$ in the z-plane corresponds to N, the north pole of the sphere, an ordinary point.

22. The transformation $w = 1/z$ in the plane corresponds to that from (X,Y,Z) to $(X,-Y,1-Z)$ on the sphere. This is a rotation of the sphere about the diameter of the sphere parallel to the x-axis through $180°$. Hence it is everywhere conformal.

23. From Problems 16 to 22 and the properties of the bilinear transformation of Eq. (6), it follows that the corresponding transformation on the sphere is one-one and conformal everywhere (even at N) in a *literal* sense. Moreover it takes circles into circles.

38. Inversion in any "circle"

Consider a given circle C with center at P_0 and radius r. As in Fig. 14, let P_1 and P_2 be any two points on the same extended radius which have $P_0P_1 \cdot P_0P_2 = r^2$. Then P_1 and P_2 are called *inverse* points with respect to the circle C. If C_{12} is any circle through P_1 and P_2, and P_0T is the tangent to this circle at T, we have

$$\overline{P_0T}^2 = P_0P_1 \cdot P_0P_2 = r^2, \qquad P_0T = r. \tag{18}$$

This shows that T is on C. Thus the tangent P_0T to C_{12} is along a radius of C, so that the circle C_{12} cuts C at right angles.

In the closed plane, we regard $P_2 = I$ or $z_2 = \infty$ as inverse to $P_1 = P_0$. This makes the "circles" through P_1 and P_2 the straight lines through the center P_0, or extended radii, which cut C at right angles.

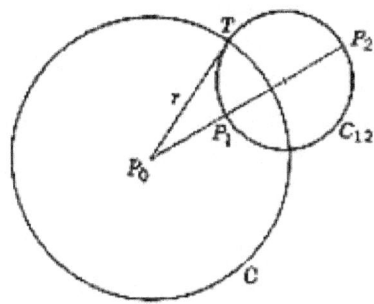

Fig. 14

If P_1 is on C, $P_2 = P_1$. Here, by taking limits as $P_1 \to C$, we interpret the "circles" through P_1 and P_2 as those through P_1 tangent to a radius.

Let P_0, P_1, P_2 be the points z_0, z_1, z_2. Then we have

$$|z_1 - z_0||z_2 - z_0| = P_0P_1 \cdot P_0P_2 = r^2. \qquad (19)$$

And, since P_1 and P_2 are on the same extended radius, we have

$$\arg(z_2 - z_0) = \arg(z_1 - z_0). \qquad (20)$$

The two Eqs. (19) and (20) combined are equivalent to

$$(z_1 - z_0)(\bar{z}_2 - \bar{z}_0) = r^2 \operatorname{cis} 0 = r^2. \qquad (21)$$

The equation of the circle C may be written

$$|z - z_0|^2 = r^2 \quad \text{or} \quad (z - z_0)(\bar{z} - \bar{z}_0) = r^2. \qquad (22)$$

When multiplied by any real constant $a \neq 0$, this becomes

$$az\bar{z} + Sz + \bar{S}\bar{z} + d = 0, \qquad (23)$$

the form given in Eq. (13). A comparison of the last three equations shows that if Eq. (21) is multiplied by a it becomes

$$az_1\bar{z}_2 + Sz_1 + \bar{S}\bar{z}_2 + d = 0. \qquad (24)$$

This is one form of the condition that z_1 and z_2 be inverse points with respect to the circle C given by Eq. (23).

Two points P_1 and P_2 are *inverse* with respect to a straight line L if P_2 is the *reflection* of P_1 in L, so that L is the perpendicular bisector of P_1P_2. Thus the "circles" through P_1 and P_2 cut L at right angles.

We regard I as the inverse of itself with respect to any straight line L, and interpret the "circles" through $P_1 = I$ and $P_2 = I$ as the straight lines perpendicular to L.

Let L be the straight line which is the perpendicular bisector of the segment joining the points $z = A$ and $z = B$. Then

$$|z - A|^2 = |z - B|^2 \quad \text{or} \quad (z - A)(\bar{z} - \bar{A}) = (z - B)(\bar{z} - \bar{B}) \quad (25)$$

is one form of the equation of the straight line. When multiplied by any real constant $q \neq 0$ this becomes

$$Sz + \bar{S}\bar{z} + d = 0, \quad (26)$$

the form given by Eq. (13) with $a = 0$.

Let z_1, z_2 be any other pair of points inverse with respect to L. Then triangle Az_1B is the reflection in L of triangle Bz_2A, whose reflection in the real axis is $\overline{Bz_2A}$. Hence triangle Az_1B is congruent to triangle $\overline{Bz_2A}$, with the sense of angles preserved, so that

$$\frac{A - z_1}{B - z_1} = \frac{\bar{B} - \bar{z}_2}{\bar{A} - \bar{z}_2}, \quad (z_1 - A)(\bar{z}_2 - \bar{A}) = (z_1 - B)(\bar{z}_2 - \bar{B}). \quad (27)$$

Conversely, if Eq. (27) holds, the triangle $\overline{Bz_2A}$ is similar to triangle Az_1B, and hence congruent to it because $|\bar{B} - \bar{A}| = |A - B|$. But for B the reflection of A in L, the congruence of triangle Bz_2A and Az_1B with reversal of sense makes z_2 the reflection of z_1 in L.

A comparison of Eqs. (25), (26), and (27) shows that when Eq. (27) is multiplied by the real factor $q \neq 0$, it becomes

$$Sz_1 + \bar{S}\bar{z}_2 + d = 0. \quad (28)$$

This is one form of the condition that z_1 and z_2 be inverse points with respect to the straight line L given by Eq. (26).

We sometimes express the fact that P_1 and P_2 are inverse with respect to a "circle" in the sense just defined by saying that P_2 is the reflection of P_1 in the "circle." We shall now prove that the transformation $w = 1/z$ takes two points z_1 and z_2 which are inverse with respect to a "circle" C into two image points w_1 and w_2 which are inverse with respect to the transformed "circle" C'. To show this, let the equation of C, in the form of Eq. (13), be

$$az\bar{z} + Sz + \bar{S}\bar{z} + d = 0. \tag{29}$$

Then by Eqs. (24) and (28), the relation

$$az_1\bar{z}_2 + Sz_1 + \bar{S}\bar{z}_2 + d = 0 \tag{30}$$

is the condition that z_1 and z_2 be inverse with respect to C. By substituting $z_1 = 1/w_1$, $\bar{z}_2 = 1/\bar{w}_2$ in this relation we find:

$$dw_1\bar{w}_2 + \bar{S}w_1 + S\bar{w}_2 + a = 0. \tag{31}$$

This is the condition that the two image points w_1 and w_2 be inverse with respect to the circle given by

$$dw\bar{w} + \bar{S}w + S\bar{w} + a = 0. \tag{32}$$

But by Eq. (16), this is the transformed "circle" C', so that w_1 and w_2 are inverse with respect to C', as was to be proved.

The figure consisting of a "circle" and two points inverse to it retains its character under any shape-preserving transformation having the form of Eq. (1), as well as for the reflection in the real axis, $W = \bar{w}$. Hence by Eq. (8) the result just proved also holds for the general bilinear transformation of Eq. (6), as well as for an inversion in the unit circle, or in any other "circle."

A family of "circles" which pass through two points is called a *pencil*. Our discussion of inverse points shows that P_1 and P_2 are inverse with respect to a "circle" C if and only if each "circle" of the pencil through P_1 and P_2 cuts C at right angles. It follows from this that any transformation which is isogonal and takes "circles" into "circles" must take inverse points with respect to C into inverse points with respect to the image "circle" C'. This gives a second proof of the theorem:

The general bilinear transformation, as well as a reflection in any "circle," takes two points which are inverse with respect to a "circle" C into two image points which are inverse with respect to the transformed "circle" C'.

39. Equations of coaxial circles

The equation of a "circle" through two distinct points z_1 and z_2, or of a "circle" with respect to which z_1 and z_2 are inverse points may be put in a simple and useful form which we proceed to develop. We first note that the bilinear transformation

$$w = \frac{z - z_1}{z - z_2} \tag{33}$$

takes z_1 into $w_1 = 0$, or O' and takes z_2 into $w_2 = \infty$, or I'.

Any "circle" through O' and I' is a straight line on which

$$\arg w = \alpha \quad \text{or} \quad \arg w = \alpha + \pi. \tag{34}$$

Each of these relations represents a ray or half-line from O' to I', the two rays together making up one straight line.

It follows from Eqs. (33) and (34) that

$$\arg \frac{z - z_1}{z - z_2} = \alpha \quad \text{or} \quad \arg \frac{z - z_1}{z - z_2} = \alpha + \pi \tag{35}$$

represents a "circle" through z_1 and z_2. The second relation represents an arc joining z_1 and z_2, whose half-tangent at z_1 has the direction such that angle $z_2z_1T_1 = \alpha$, while the half-tangent at z_2 has the direction such that angle $z_1z_2T_2 = -\alpha$. Each "circle" through z_1 and z_2 is represented by Eq. (35) for just one value of α such that $0 \leq \alpha < \pi$, $\pi \leq \alpha + \pi < 2\pi$. Collectively, these "circles" make up a system of *coaxial circles* of the *common point* kind, with z_1 and z_2 as the common points.

Any circle with respect to which O' and I' are inverse points is a circle on which $|w| = a$. It follows from Eq. (33) that

$$\left| \frac{z - z_1}{z - z_2} \right| = a \tag{36}$$

is a circle with respect to which z_1 and z_2 are inverse. Each such circle is represented by Eq. (36) for just one positive value of a. Collectively, these "circles" make up a system of *coaxial circles* of the *limiting point* kind, with z_1 and z_2 as the limiting points.

Since z_1 and z_2 are inverse with respect to any circle of Eq. (36), any circle through z_1 and z_2 given by Eq. (35) must cut this circle at right angles. This of course also follows from the orthogonality of their images in the w-plane. This proves:

For the two systems of coaxial circles determined by two points, Eqs. (35) and (36), each circle of either system cuts every circle of the other system at right angles.

Exercise 21

1. Show that the point z_2, inverse to z_1 in the "circle" given by $az\bar{z} + Sz + \bar{S}\bar{z} + d = 0$ is

$$z_2 = -\frac{\bar{S}z_1 + d}{a\bar{z}_1 + S}.$$

2. From Eq. (22) deduce that the point z_2 inverse to z_1 in the circle with radius r and center at z_0 is

$$z_2 = z_0 + \frac{r^2}{\bar{z}_1 - \bar{z}_0}.$$

3. By Problem 6 of Exercise 20, $\operatorname{Re} z\bar{z}_0 = b$ is the equation of a straight line perpendicular to the direction of z_0. Show that if z_2 is the reflection of z_1 in this straight line,

$$z_2 = \frac{2b - z_0\bar{z}_1}{\bar{z}_0}.$$

4. Verify that $b/\bar{z}_0$ is on the straight line $\operatorname{Re} z\bar{z}_0 = b$ of Problem 3. The reflection of $A = 0$ in this straight line is $B = 2b/\bar{z}_0$, since $B - A = 2b/\bar{z}_0 = (2b/z_0\bar{z}_0)z_0$ which has the normal direction of z_0 when $b \neq 0$. Use these values and Eq. (27) to check the result of Problem 3 for the case $b \neq 0$. To check the case $b = 0$, take $A = z_0$, $B = -z_0$.

5. By Problem 3 of Exercise 20, if

$$w = \frac{Az + B}{Cz + D}, \quad z = \frac{-Dw + B}{Cw - A}, \quad \text{so that } \bar{z} = \frac{-\bar{D}\bar{w} + \bar{B}}{\bar{C}\bar{w} - \bar{A}}.$$

Deduce from this that the image of the "circle" given by $az\bar{z} + Sz + \bar{S}\bar{z} + d = 0$ is given by $a'w\bar{w} + S'w + \bar{S}'\bar{w} + d' = 0$, $a' = aA\bar{A} + SA\bar{C} + \bar{S}\bar{A}C + dC\bar{C}$ and $d' = aB\bar{B} + SB\bar{D} + \bar{S}\bar{B}D + dD\bar{D}$ are real, while $S' = aA\bar{B} + SA\bar{D} + \bar{S}\bar{C}B + dC\bar{D}$. Note that $S'\bar{S}' - a'd' =$

$(AD - BC)(\overline{AD} - \overline{BC})(S\overline{S} - ad) = |AD - BC|^2(S\overline{S} - ad)$ preserves its nonzero character. This proves again that under Eq. (6), "circles" go into "circles."

6. By a calculation like that of Problem 5 verify that the inverse points z_1 and z_2 with respect to the first "circle" which satisfy $az_1\bar{z}_2 + Sz_1 + \overline{S}\bar{z}_2 + d = 0$ go into points which satisfy $a'w_1\bar{w}_2 + S'w_1 + \overline{S}'\bar{w}_2 + d = 0$. This again proves that w_1 and w_2 are inverse points with respect to the transformed "circle."

7. Verify that if w, w_1 are the images of z, z_1 under the transformation of Eq. (6),
$$w - w_1 = \frac{AD - BC}{(Cz + D)(Cz_1 + D)}(z - z_1).$$

8. From Problem 7 deduce that
$$\frac{w - w_1}{w - w_2} = \frac{Cz_1 + D}{Cz_2 + D}\frac{z - z_1}{z - z_2}.$$

Use this result to show that the system of coaxial "circles" of Eqs. (35) or (36) related to z_1 and z_2 goes into the coaxial system of the same kind related to w_1 and w_2.

The *cross ratio* of four points taken in the order
$$z_1, z_2, z_3, z_4 \text{ is } \frac{(z_1 - z_4)(z_3 - z_2)}{(z_1 - z_2)(z_3 - z_4)},$$
or the simple ratio resulting from taking the limit if just one of the points is I or $z = \infty$.

9. Prove that four points are *concyclic*, that is, lie on a "circle" if and only if their cross ratio is real. *Hint:* In Eq. (35) replace z_1 by z_4. Then replace z first by z_1 and then by z_2.

10. Prove that the cross ratio is invariant under the general bilinear transformation of Eq. (6), that is,
$$\frac{(w_1 - w_4)(w_3 - w_2)}{(w_1 - w_2)(w_3 - w_4)} = \frac{(z_1 - z_4)(z_3 - z_2)}{(z_1 - z_2)(z_3 - z_4)}.$$

Hint: In Problem 8 replace z_1, w_1 by z_4, w_4. Then replace z, w first by z_1, w_1 and then by z_3, w_3.

11. Prove that there is one and only one bilinear transformation (6) which takes any three distinct points z_1, z_2, z_3 in the closed z-plane into $w_1 = 0, w_2 = 1, w_3 = \infty$ or I' in the closed w-plane, namely,

$$w = \frac{(z_1 - z)(z_3 - z_2)}{(z_1 - z_3)(z_2 - z)}.$$

Hint: Use Problem 10, with z_4, w_4 replaced by z, w to show that the relation must hold, and notice that here

$$AD - BC = (z_1 - z_2)(z_2 - z_3)(z_3 - z_1) \neq 0$$

since the points are distinct.

12. Prove that there is one and only one bilinear transformation (6) which takes any three distinct points z_1, z_2, z_3 in the closed z-plane into any three distinct points w_1, w_2, w_3 in the closed w-plane, which may be found by solving the following relation for w:

$$\frac{(w_1 - w)(w_3 - w_2)}{(w_1 - w_3)(w_2 - w)} = \frac{(z_1 - z)(z_3 - z_2)}{(z_1 - z_3)(z_2 - z)}.$$

Hint: The relation must hold by Problem 10. Let each function equal W. Then T_1, the transformation from z to W as well as T_2, the transformations from w to W each have the form given in Problem 11, with $AD - BC \neq 0$. Hence by Problems 3 and 4 of Exercise 20, the solved form of $T_2 w = T_1 z$, or $w = T_2^{-1} T_1 z$ leads to a bilinear transformation $T_2^{-1} T_1$ with determinant different from zero.

13. Use Problems 9 and 10 to prove, once again, that the bilinear transformation (6) takes "circles" into "circles."

14. Verify that the *fixed points*, or points such that $w = z$, of the transformation (6) are the roots of the quadratic equation $Cz^2 + (D - A)z - B = 0$.

15. Let $(D - A)^2 + 4BC \neq 0$, so that in Problem 14 there are two distinct fixed points, P and Q. Deduce from Problem 8 that if $C \neq 0$, the transformation is equivalent to

$$\frac{w - P}{w - Q} = R \frac{z - P}{z - Q},$$

where $R \neq 0$ is a suitable complex constant. If $C = 0$, then $Q = \infty$ and $w - P = R(z - P)$.

16. In Problem 14 let $(D - A)^2 + 4BC = 0$. Then if $B = C = 0$, $w = z$, the identity with all points fixed. If $C = 0, B \neq 0$, $w = z + B$, the translation of Eq. (3), with one fixed point at infinity. And if $C \neq 0$, by putting

$$Z = \frac{1}{z - P}, \quad W = \frac{1}{w - P},$$

deduce that the transformation is equivalent to

$$\frac{1}{w-P} = \frac{1}{z-P} + R,$$

where P is the finite fixed point, and $R \neq 0$ is a suitable complex constant.

40. Particular bilinear transformations

It was shown in Problem 12 of Exercise 21 that the unique bilinear transformation which takes any three distinct points in z_1, z_2, z_3 in the closed z-plane into any three distinct points w_1, w_2, w_3 in the closed w-plane may be obtained by equating the *cross ratios*,

$$\frac{(w_1 - w)(w_3 - w_2)}{(w_1 - w_2)(w_3 - w)} = \frac{(z_1 - z)(z_3 - z_2)}{(z_1 - z_2)(z_3 - z)}. \tag{37}$$

In particular, if $w_1 = 0$ and $w_3 = \infty$, this may be reduced to

$$w = \frac{w_2(z_2 - z_3)}{z_2 - z_1} \frac{z - z_1}{z - z_3}. \tag{38}$$

If neither z_1 nor z_3 is infinite, we may write

$$w = R \frac{z - z_1}{z - z_3}. \tag{39}$$

This form, where $R \neq 0$ is a complex constant, is often useful.

The "circle" through z_1, z_2, z_3 is *oriented* by the order of these points. We regard the oriented "circle" as "bounding" the region on our left as we traverse the "circle" in this order. Thus a circle traversed in the positive direction "bounds" its interior, while traversed in the negative direction it "bounds" its exterior. And a straight line "bounds" the half-plane generated by rotating the ray from any point on it in the positive direction through a positive angle less than π. The result stated in Section 32 on the determination of the conformal map of a finite region R by the mapping of its boundary can be extended to infinite regions as well if the transformation is bilinear, because the mapping under a bilinear transformation is conformal at I and I'. Hence the region "bounded" by the oriented "circle" through z_1, z_2, z_3 is mapped on the region "bounded" by the oriented "circle" through w_1, w_2, w_3 by the transformation of Eq. (37). This gives one solution of the problem of mapping the interior or exterior of any circle, or any half-plane on any other "circular" region by a bilinear transformation. To illustrate how a judicious choice of the data simplifies the solution, we shall solve a few specific examples in detail.

Example 1. Find all bilinear transformations which map the half-plane Im $z = y > 0$ on the interior of the unit circle, $|w| < 1$.

Solution: To use Eq. (38) or (39) we require z_1 whose image is $w_1 = 0$, and z_3 whose image is $w_3 = \infty$. Since w_1 is inside the unit circle, it will be the image of some finite point in the half-plane Im $z > 0$, say $z_1 = z_0$ with Im $z_0 > 0$. Then $w_3 = \infty$, the inverse of w_1 in the unit circle, will correspond to $\bar{z}_0$, the reflection of $z_1 = z_0$ in $y = 0$, the boundary of the half-plane. If in Eq. (38) $z_2 = \infty$ or I, in Eq. (39) $R = w_2$. Since I is on the straight line $y = 0$, its image w_2 must be some point on the unit circle. If we call its argument θ, then $w_2 = 1 \operatorname{cis} \theta = e^{i\theta}$. Hence in Eq. (39) $R = e^{i\theta}$ and the required transformation must be

$$w = e^{i\theta} \frac{z - z_0}{z - \bar{z}_0}, \qquad \operatorname{Im} z_0 > 0. \tag{40}$$

Example 2. Find all bilinear transformations which map the interior of the unit circle $|z| < 1$ onto the interior of the unit circle $|w| < 1$.

Solution: Since $w = 0$ is inside $|w| < 1$, it will be the image of some point inside $|z| < 1$, say z_0 with $|z_0| < 1$. Then $w_3 = \infty$, the inverse of $w_1 = 0$ in $|w| = 1$, will correspond to $z_3 = 1/\bar{z}_0$, the inverse of $z_1 = z_0$ in $|z| = 1$. Hence by Eq. (39) we have

$$w = R \frac{z - z_0}{z - 1/\bar{z}_0} = R\bar{z}_0 \frac{z - z_0}{\bar{z}_0 z - 1}. \tag{41}$$

This shows that if w_2 is the image of $z_2 = 1$,

$$w_2 = R\bar{z}_0 \frac{1 - z_0}{\bar{z}_0 - 1}, \qquad R\bar{z}_0 = -w_2 \frac{\bar{z}_0 - 1}{z_0 - 1}. \tag{42}$$

Since $z_2 = 1$ is on $|z| = 1$, w_2 is on $|w| = 1$ and hence $|w_2| = 1$. Also $\bar{z}_0 - 1$ is the conjugate of $z_0 - 1$ so that $|\bar{z}_0 - 1| = |z_0 - 1|$. Using these results, and taking absolute values in Eq. (42) shows that $|R\bar{z}_0| = 1$. And if we let arg $R\bar{z}_0 = \theta$, $R\bar{z}_0 = e^{i\theta}$. Thus Eq. (41) becomes

$$w = e^{i\theta}\frac{z - z_0}{\bar{z}_0 z - 1}, \qquad |z_0| < 1 \tag{43}$$

which is the required transformation.

We note that Eq. (43) gives a family of transformations which take any point z_0 inside $|z| = 1$ into 0, the center of $|w| = 1$. In particular, if $z_0 = 0$, $w = -e^{i\theta}z = e^{i\alpha}z$ represents the rotations about the origin. These include the identity, $w = z$, for $\theta = \pi$ or $\alpha = 0$.

Exercise 22

In each case verify that the given domain D is "bounded" by the oriented "circle" through z_1, z_2, z_3, while D' is "bounded" by that through w_1, w_2, w_3. And use Eq. (37) to find the bilinear transformation which maps z_1, z_2, z_3 onto w_1, w_2, w_3, respectively, and which in consequence maps D onto D'.

1. $D: |z| < 1$, $z_1 = 1$, $z_2 = i$, $z_3 = -1$. $D': |w| < 1$, $w_1 = i$, $w_2 = -1$, $w_3 = -i$. Ans. $w = iz$. This is Eq. (43) with $z_0 = 0$, $\theta = -\pi/2$.

2. $D: |z| < 1$, $z_1 = 1$, $z_2 = i$, $z_3 = -1$. $D': |w| > 1$, $w_1 = i$, $w_2 = 1$, $w_3 = -i$. Ans. $w = i/z$.

3. $\text{Im } z = y > 0, z_1 = -1, z_2 = 0, z_3 = 1$. $D': |w| < 1, w_1 = -i, w_2 = 1, w_3 = i$. Ans. $w = (-z + i)/(z + i)$. This is Eq. (40) with $z_0 = i, \theta = \pi$.

4. $D: \text{Re } z = x > 0, z_1 = i, z_2 = 0, z_3 = -i$. $D': |w| < 1, w_1 = i, w_2 = -1, w_3 = -i$. Ans. $w = (z - 1)/(z + 1)$.

5. $D: |z| > 1$, $z_1 = 1$, $z_2 = -i$, $z_3 = -1$. $D': \text{Re } w = u > 0$, $w_1 = i, w_2 = 0, w_3 = -i$. Ans. $w = (z + i)/(z - i)$.

6. Find the bilinear transformation which maps $D: |z| < 1$ on $D': |w - 1| < 1$, takes the interior point (center) $z_1 = 0$ into the interior point $w_1 = \frac{1}{2}$, and takes the boundary point $z_2 = 1$ into the boundary point $w_2 = 0$. Hint: $z_3 = \infty$, inverse to $z_1 = 0$ in $|z| = 1$ must go into $w_3 = -1$, inverse to $w = \frac{1}{2}$ in $|w - 1| = 1$. Ans. $w = (-z + 1)/(z + 2)$.

7. Show that every transformation

$$w = \frac{Az + B}{\bar{B}z + \bar{A}} \quad \text{with } |A| > |B|,$$

is equivalent to Eq. (43) with $z_0 = -B/A$ and $\theta = \pi + 2 \arg A$.

8. Show that the most general bilinear transformation which maps the upper half-plane $\text{Im } z = y > 0$ on the upper half-plane $\text{Im } w = v > 0$ has one of the three forms

$$w = q\frac{z - x_1}{z - x_2} \quad \text{with } q(x_1 - x_3) > 0,$$

$$w = \frac{-p}{z - x_2} \quad \text{with } p > 0,$$

or

$$w = p(z - x_1) \quad \text{with } p > 0.$$

Also show that every transformation $w = \dfrac{az + b}{cz + d}$ with real coefficients such that $ad - bc > 0$ is equivalent to one of these forms.

9. Reasoning as in Example 2, show that the most general bilinear transformation which maps the interior of the circle $|z| < r$ onto the interior of the unit circle $|w| < 1$ is

$$w = re^{i\theta}\frac{z - z_0}{\bar{z}_0 z - r^2} \quad |z_0| < r.$$

10. Three circles each pass through the point z_0. Taken in pairs they determine three other points of intersection z_1, z_2, z_3. The circular arcs z_2z_3, z_3z_1, z_1z_2 not containing z_0 are the sides of a curvilinear triangle D. Find the transformation which maps D on a triangle with straight sides taking z_1 into $w_1 = 0$ and z_2 into $w_2 = 1$.

Ans. $w = \dfrac{z_2 - z_0}{z_2 - z_1}\dfrac{z - z_1}{z - z_0}$.

11. Find the bilinear transformation which maps the part of the right half-plane, $\text{Re } z = x > 0$, which is outside the circle $|z - a| > r$ where $0 < r < a$, on the area between two concentric circles each with center at $w = 0$ in the w-plane and such that $z = 0$ on $x = 0$ goes into $w = -1$ on the larger circle. *Hint:* The points $\pm b$ are inverse with respect to $x = 0$, and with respect to $|z - a| = r$ if $b = \sqrt{a^2 - r^2}$. Ans. $w = (z - b)/(z + b)$. The radius of the larger concentric circle is 1, that of the smaller is $\sqrt{a - b}/\sqrt{a + b}$.

12. Let the two circles $C_1: |z - z_1| = r_1$ and $C_2: |z - z_2| = r_2$ with $z_1 \neq z_2$ have no common points. Show that the domain D which has both C_1 and C_2 as its boundary can be mapped by a bilinear transformation onto the domain $r < |w| < Kr$ for a suitable value of the

constant K. *Hint:* There are two points z_3 and z_4 which are inverse with respect to both C_1 and C_2, whose existence may be shown by putting $W = 1/(z - z_1 - r_1)$, and noting that this takes D into a domain essentially that of Problem 11. Then the required transformation has the form $w = A\,(z - z_3)/(z - z_4)$, and K is the ratio of the larger to the smaller of p and q, where $\left|\dfrac{z - z_3}{z - z_4}\right| = p$ on C_1, and $= q$ on C_2.

7

INTEGRAL THEOREMS

This chapter is an introduction to the complex integral calculus. It begins with a review of the definition of definite integral as given in the calculus for real functions. The concept of definite integral is then extended to the complex domain. Here the limits of integration may be any two points in the complex plane. And the path of integration may be any sufficiently smooth curve joining these points. This definite integral is seen to have linearity properties both with respect to the function used as an integrand, and with respect to the path of integration. We show that if the integrand is the derivative of an analytic function and the path is suitably restricted there is an analog of the fundamental theorem of the calculus. We then prove Cauchy's integral theorem, as well as the stronger form due to Goursat, on the vanishing of the definite integral of an analytic function over several suitably restricted closed paths.

This enables us to prove Cauchy's integral formula which expresses $f(z)$ as a contour integral around a closed contour surrounding the point z. The integral formula is useful in proving a number of fundamental properties of analytic functions, such as their possession of derivatives of all orders and their expansibility in Taylor's series, as we shall show in Chapter 8.

41. Real definite integrals

Let $\phi(t)$ be a real continuous function of the real variable t. Then for $t' < t''$, the *definite integral* of $\phi(t)$ from the lower limit t' to the upper limit t'' may be defined by the following procedure.

First, divide the interval $t' \leq t \leq t''$ into n parts by values t_k such that

$$t' = t_0 < t_1 < t_2 < \ldots < t_n = t''. \tag{1}$$

Then select an intermediate point $\bar{t}_k$ in each subinterval

$$t_{k-1} \leq \bar{t}_k \leq t_k. \tag{2}$$

Call the length of the subinterval Δt_k, so that

$$\Delta t_k = t_k - t_{k-1}. \tag{3}$$

And for each choice of n we form a sum

$$S_n(t_k, \bar{t}_k) = \sum_{k=1}^{n} \phi(\bar{t}_k) \Delta t_k. \tag{4}$$

As the notation $S_n(t_k,\bar{t}_k)$ suggests, each sum depends on the choice of n, and then on the further choice of t_k and $\bar{t}_k$ for $1 \leq k \leq n - 1$.

For a given $S_n(t_k,\bar{t}_k)$, call the largest length of any subinterval δ_n, so that $\delta_n = \max \Delta t_k$. Now consider an infinite sequence of sums for which $n \to \infty$ and $\delta_n \to 0$. Then, as shown in calculus, the sequence approaches a limit. This limit is the same for all sequences, and it is independent of the choices of t_k and $\bar{t}_k$ except for the one restriction that $\delta_n \to 0$. The common limit is denoted by $\int_{t'}^{t''} \phi(t)\, dt$. Thus

$$\int_{t'}^{t''} \phi(t)\, dt = \lim_{n \to \infty} \sum_{k=1}^{n} \phi(\bar{t}_k) \Delta t_k, \quad \max \Delta t_k \to 0. \tag{5}$$

As a suggestive abbreviated summary of the discussion just given, Eq. (5) defines the *definite integral* of $\phi(t)$ from t' to t''.

Our discussion shows that the definite integral, $\int_{t'}^{t''} \phi(t)\, dt$, is a number determined by the values of $\phi(t)$ for t in the interval $t' \leq t \leq t''$. These values of t correspond to points in the z-plane which make up the segment of the real axis which joins $z' = t'$ to $z'' = t''$. Hence we may say that the integration in Eq. (5) is carried out along the real axis.

42. Curved paths of integration

We wish to extend the concept of definite integration to the complex plane. The limits of integration z', z'' may be any two complex numbers. In this section we assume that they are finite and distinct. To select intermediate values z_k and $\tilde{z}_k$, we must specify a *path*, or curve, C joining the two points z' and z''. We assume that C is smooth. Then C may be given a parametric representation

$$x = x(t), y = y(t), \quad t' \leq t \leq t'', \tag{6}$$

where each of the functions $x(t)$, $y(t)$ is single-valued and has a continuous derivative of the first order. Under these conditions, the path C has a finite total arc length L given by

$$L = \int_{t'}^{t''} \sqrt{[x'(t)]^2 + [y'(t)]^2}\, dt. \tag{7}$$

Since $z = x + iy$, $z(t) = x(t) + iy(t)$. And the relation

$$z = z(t), \quad t' \leq t \leq t'' \tag{8}$$

is equivalent to Eq. (6). In particular, the end points are

$$z' = z(t'), \quad z'' = z(t''). \tag{9}$$

The length of arc along C from z' to $z = z(t)$ is

$$s = s(t) = \int_{t'}^{t} \sqrt{[x'(t)]^2 + [y'(t)]^2}\, dt. \tag{10}$$

This has a derivative given by

$$\frac{ds}{dt} = s'(t) = \sqrt{[x'(t)]^2 + [y'(t)]^2}. \tag{11}$$

Hence ds/dt will never be zero on C if and only if

$$[x'(t)]^2 + [y'(t)]^2 > 0 \quad \text{for} \quad t' \leq t \leq t''. \tag{12}$$

When this is the case, we call t a *proper* parameter. The length s in Eq. (10) is one possible proper parameter, since if $t = s$, $ds/dt = 1 \neq 0$. Let us assume that the t in Eq. (6) is a proper parameter, so that Eq. (12) holds.

From Eq. (11), $s'(t)$ is a real and continuous function of t on the closed interval $t' \leq t \leq t''$. Hence it must assume its maximum value M, and its minimum value m, in this closed interval. The condition (12) makes $m > 0$, so that

$$0 < m \leq s'(t) \leq M \quad \text{for} \quad t' \leq t \leq t''. \tag{13}$$

For the t_k of Eq. (1), let $s_k = s(t_k)$. As in Eq. (3), let

$$\Delta t_k = t_k - t_{k-1}, \quad \Delta s_k = s_k - s_{k-1}. \tag{14}$$

Then by the mean value theorem of differential calculus we have for some suitably chosen t_k^* such that $t_{k-1} < t_k^* < t_k$,

$$\frac{\Delta s_k}{\Delta t_k} = s'(t_k^*), \quad \text{and} \quad 0 < m \leq \frac{\Delta s_k}{\Delta t_k} \leq M \tag{15}$$

by Eq. (13). Since $\Delta t_k > 0$, this implies that

$$\Delta t_k \leq \frac{1}{m} \Delta s_k, \quad \Delta s_k \leq M \Delta t_k. \tag{16}$$

The first relation combined with Eq. (1) shows that

$$0 = s' = s_0 < s_1 < s_2 < \ldots < s_n = s'' = L. \tag{17}$$

Also, if $\bar{s}_k = s(\bar{t}_k)$, and Eq. (2) holds, then

$$s_{k-1} \leq \bar{s}_k \leq s_k. \tag{18}$$

Since $\Delta t_k > 0$ makes $\Delta s_k > 0$, the t to s relation given by $s(t)$ is one-one. If we start with values s_k satisfying Eq. (17) and $\bar{s}_k$ satisfying Eq. (18), the corresponding values t_k and $\bar{t}_k$ must satisfy Eqs. (1) and (2), because of the second relation of Eq. (16). The conclusions of this paragraph are all immediate consequences of the fact that $s(t)$ is a strictly monotonic function, or that $s'(t) > 0$, which is the characteristic property of a proper parameter.

Consider next a sequence of increasing values of n. For each n, we select corresponding values t_k and $s_k = s(t_k)$ satisfying the relations (1) and (17). As just remarked, we may start with either set. Then the first relation of Eq. (16) shows that

$$\max \Delta t_k \to 0 \quad \text{if } \max \Delta s_k \to 0 \text{ as } n \to \infty. \tag{19}$$

And conversely, by the second relation of Eq. (16).

43. Integrals along curves

We are now ready to discuss the integration of a function $f(z)$ along a curved path. Let $f(z)$ be a function of a complex variable which is continuous at each point of the curve C of Eq. (8). Then the *definite integral* of $f(z)$ from z' to z'' along C is defined by the following procedure.

First, divide C into n parts in any manner. Denote the points of subdivision, taken in order, by

$$z_0 = z', \quad z_1, \quad z_2, \quad \ldots, \quad z_{k-1}, \quad z_k, \quad \ldots, \quad z_n = z''. \tag{20}$$

On each of the subarcs $z_{k-1}z_k$ choose a point $\bar{z}_k$. Call the complex number representing the directed chord of the subarc Δz_k, so that

$$\Delta z_k = z_k - z_{k-1}. \tag{21}$$

And for each choice of n form a sum

$$S_n(z_k, \bar{z}_k) = \sum_{k=1}^{n} f(\bar{z}_k) \Delta z_k. \tag{22}$$

As the notation $S_n(z_k, \bar{z}_k)$ suggests, each sum depends on the choice of n, and then on the further choice of z_k and $\bar{z}_k$ for $1 \leq k \leq n - 1$.

Each point z on the curve C determines s, the length of arc along C from z' to z, and since the parameter in Eq. (8) is proper, there is just one value of t such that $s = s(t)$, $z = z(t)$. We note that the points $z_k, \bar{z}_k$ correspond in this way to values $s_k, \bar{s}_k$ which satisfy Eqs. (17) and (18), and hence to values $t_k, \bar{t}_k$ which satisfy Eqs. (1) and (2). Also the length of the subarc $z_{k-1}z_k$ is $s_k - s_{k-1} = \Delta s_k$.

For a given $S_n(z_k, \bar{z}_k)$, let δ_n denote the largest length of any subarc, $z_{k-1}z_k$, so that $\delta_n = \max \Delta s_k$. Now consider an infinite sequence of sums for which $n \to \infty$ and $\delta_n \to 0$. Then, as we shall show presently, the sequence approaches a limit. This limit is the same for all sequences, and it is independent of the choices of z_k and $\bar{z}_k$ on C except for the one restriction that $\delta_n \to 0$. The common limit is denoted by $\int_{z'}^{z''} f(z)\, dz$ or $\int_C f(z)\, dz$. Thus

$$\int_C f(z)\, dz = \lim_{n \to \infty} \sum_{k=1}^{n} f(\bar{z}_k) \Delta z_k, \quad \max \Delta s_k \to 0. \tag{23}$$

As a suggestive summary of the procedure just described, Eq. (23) defines the *integral of $f(z)$ along the path C*.

Except for affixes, the typical term in Eq. (22) is $f(z) \Delta z$. To express this as a combination of real terms we write

$$z = x + iy, \quad \Delta z = \Delta x + i \Delta y, \quad f(z) = u + iv. \tag{24}$$

We then obtain the following decomposition

$$f(z)\,\Delta z = (u + iv)(\Delta x + i\,\Delta y)$$
$$= (u\,\Delta x - v\,\Delta y) + i(u\,\Delta y + v\,\Delta x). \tag{25}$$

This may be used to express $f(z_k)\,\Delta z_k$ as a combination of four real terms, of which the first is $u(\bar{x}_k,\bar{y}_k)\,\Delta x_k$. By the mean value theorem, for a suitably chosen t_k^* such that

$$t_{k-1} \leq t_k^* \leq t_k, \quad \Delta x_k = x'(t_k^*)\,\Delta t_k. \tag{26}$$

Thus the sum obtained from the first term $u(\bar{x}_k,\bar{y}_k)\,\Delta x_k$ is

$$\sum_{k=1}^{n} u(\bar{x}_k,\bar{y}_k)\,\Delta x_k = \sum_{k=1}^{n} u[x(\bar{t}_k), y(\bar{t}_k)]x'(t_k^*)\,\Delta t_k. \tag{27}$$

The sum on the right contains two intermediate values, $\bar{t}_k$ and t_k^* in place of the one used in Eq. (4). However, since all the functions involved are continuous, we may apply Duhamel's principle. Compare Problem 1 of Exercise 23. This asserts that any sequence of sums with $\max \Delta t_k \to 0$ approaches the same definite integral as a similar sequence with $t_k^* = \bar{t}_k$.

The sum on the left of Eq. (27) is an approximating sum for the real line integral of $u(x,y)\,dx$ along C. This is denoted by $\int_C u(x,y)\,dx$ and is defined to be the limit of any sequence of approximating sums for which $\max \Delta s_k \to 0$. It follows from Eqs. (19) and (27) that

$$\int_C u(x,y)\,dx = \int_{t'}^{t''} u[x(t),y(t)]x'(t)\,dt. \tag{28}$$

This shows that the sequences in question have a common limit, and that this limit may be calculated from Eq. (28). It follows from the substitution rule for definite integrals that Eq. (28) holds even when t is not a proper parameter.

Each of the remaining three terms in Eq. (25) leads to a similar real line integral. Consequently, the limit in Eq. (23) exists and
$\int_C f(z)\,dz =$

$$\int_C u(x,y)\,dx - \int_C v(x,y)\,dy + i\int_C u(x,y)\,dy + i\int_C v(x,y)\,dx$$
$$= \int_C (u\,dx - v\,dy) + i\int_C (v\,dx + u\,dy). \tag{29}$$

The last expression is easily recalled by the formal separation of
$$f(z)\, dz = (u + iv)(dx + i\, dy). \tag{30}$$

44. Linear properties of integrals

For a fixed path of integration, the integral depends linearly on the function. That is, for any complex constant A,
$$\int_C Af(z)\, dz = A \int_C f(z)\, dz. \tag{31}$$
And if $g(z)$ is a second function continuous on the path C,
$$\int_C [f(z) + g(z)]\, dz = \int_C f(z)\, dz + \int_C g(z)\, dz. \tag{32}$$
These relations follow from the linear character of the sum in Eq. (22) from which the integral was obtained by a limiting process.

Let Z be a point on C distinct from the end points z' and z''. If C_1 is the arc of C from z' to Z, and C_2 is the arc of C from Z to z'', it is natural to write $C = C_1 + C_2$. Then we have
$$\int_{C_1} f(z)\, dz + \int_{C_2} f(z)\, dz = \int_{C_1 + C_2} f(z)\, dz, \tag{33}$$
since we may evaluate the integral on the right by using a sequence of sums for each of which the point Z is one of the z_k.

For a path C made up of two smooth parts C_1 and C_2, but having a corner at Z, we may use Eq. (33) to define the integral over $C_1 + C_2$. By repeated use of this relation, we may define the integral over any *contour*, such as a broken line, made up of a finite number of smooth pieces.

We may now allow z'' to coincide with z', taking as the path any *closed contour* such as a polygon. For a closed contour smooth at all points, we define the integral by Eq. (33) with any point on the contour distinct from $z'' = z'$ as Z. And we admit as a closed contour from z' back to z' the null circuit consisting of a single point. We denote it by 0, and define
$$\int_0 f(z)\, dz = 0. \tag{34}$$
This convention makes Eq. (33) continue to hold when $Z = z'$ or $Z = z''$.

Let $-C$ denote the arc of C traversed from z'' to z'. Then each Δz_k in Eq. (23) must be replaced by its negative, since the z_k of Eq. (20) are now taken in reversed order. It follows that

$$\int_{-C} f(z)\, dz = -\int_{C} f(z)\, dz. \tag{35}$$

The Eqs. (32) to (34) show that for any three points z', z'', z''',

$$\int_{z'}^{z''} f(z)\, dz + \int_{z''}^{z'''} f(z)\, dz + \int_{z'''}^{z'} f(z)\, dz = 0, \tag{36}$$

provided that the three paths of integration combine to form the null circuit,

$$\text{arc } z'z'' + \text{arc } z''z''' + \text{arc } z'''z' = 0. \tag{37}$$

Exercise 23

1. Let $\phi(x,y)$ be a real function of two real variables, continuous for x,y in the square $t' \leq x \leq t''$, $t' \leq y \leq t''$. Form the sums

$$S_n(t_k, \bar{t}_k, t_k^*) = \sum_{k=1}^{n} \phi(\bar{t}_k, t_k^*)\, \Delta t_k$$

with t_k as in Eq. (1), and intermediate points $\bar{t}_k$ and t_k^* such that $t_{k-1} \leq \bar{t}_k \leq t_k$, $t_{k-1} \leq t_k^* \leq t_k$. Prove that

$$\lim_{n \to \infty} S_n(t_k, \bar{t}_k, t_k^*) = \int_{t'}^{t''} \phi(t,t)\, dt.$$

This is one example of *Duhamel's principle*. *Hint:* Being continuous in a closed region, $\phi(x,y)$ is uniformly continuous, so that for any positive ϵ, $|\phi(\bar{t}_k, t_k^*) - \phi(\bar{t}_k, \bar{t}_k)| < \epsilon$ if $\max \Delta t_k < \delta_\epsilon$. Hence

$$|S_n(t_k, \bar{t}_k, t_k^*) - S_n(t_k, \bar{t}_k, \bar{t}_k)| < (t'' - t')\, \epsilon.$$

Each of the following is a continuously differentiable representation of the straight segment from $(0,0)$ to $(2,0)$. Verify that Eq. (10) makes $s = x$ and deduce the stated properties.

2. $x = t$, $y = 0$, $0 \leq t \leq 2$. Proper.
3. $x = 2t^2$, $y = 0$, $0 \leq t \leq 1$. One-one, not proper at $t = 0$.
4. $x = \begin{cases} \sin t, & 0 \leq t \leq \pi/2, \\ 1, & \pi/2 < t < \pi, \\ 2 + \cos t, & \pi \leq t \leq 3\pi/2, \end{cases}$ $y = 0$, $0 \leq t \leq 3\pi/2$.

Many-one and not proper for $\pi/2 \leq t \leq \pi$, that is for $(1,0)$.

Sec. 45 INTEGRAL THEOREMS 129

We define three contours from $z' = 0$ to $z'' = 1 + i$.

C_1: $x = t$, $y = t$, $0 \leq t \leq 1$, the straight-line path.
C_2: $x = 1 - \cos t$, $y = \sin t$, $0 \leq t \leq \pi/2$, a quadrant of a circle.
C_3: $x = t$, $y = 0$ for $0 \leq t \leq 1$ and $x = 1$, $y = t - 1$ for $1 \leq t \leq 2$, a broken-line path.

Check each of the following integrations.

5. $\int_{C_1} x\, dz = (1 + i)/2$, $\int_{C_2} x\, dz = \frac{1}{2} + i(1 - \pi/4)$,
 $\int_{C_3} x\, dz = \frac{1}{2} + i$.

6. $\int_{C_1} y\, dz = (1 + i)/2$, $\int_{C_2} y\, dz = \pi/4 + i/2$, $\int_{C_3} y\, dz = i/2$.

7. $\int_{C_1} z\, dz = \int_{C_2} z\, dz = \int_{C_3} z\, dz = i$.

8. $\int_{C_1} \bar{z}\, dz = 1$, $\int_{C_2} \bar{z}\, dz = 1 + i(1 - \pi/2)$, $\int_{C_3} \bar{z}\, dz = 1 + i$.

9. Let $\int_C f(z)\, |dz|$ denote the limit in Eq. (23) when we replace Δz_k by $|\Delta z_k|$. Show that the limit exists and that

$$\int_C f(z)\, |dz| = \int_C u(x,y)\, ds + i \int_C v(x,y)\, ds.$$

10. With the notation of Problem 9, verify in particular that

$$\int_C |f(z)|\, |dz| = \int_C \sqrt{u^2 + v^2}\, ds, \quad \text{and that} \int_C |dz| = L.$$

11. If the curve C is *simple*, that is, does not cross itself, in Eq. (23) we may replace the requirement that max $\Delta s_k \to 0$ by max $|\Delta z_k| \to 0$. This is because $\Delta s_k/|\Delta z_k|$ is a continuous function of the end points on C and so has a maximum, M_1, so that $\Delta s_k \leq M_1 |\Delta z_k|$. But if the curve crosses itself at Z with a single loop, if we required only max $|\Delta z_k| \to 0$, we could omit the loop entirely and so only get the integral over $z'Z$ plus that over Zz''. Check this by plotting the path with proper parameter

$$x = (1 + 2\cos t)\cos t, \quad y = (1 + 2\cos t)\sin t$$

for $0 \leq t \leq 3\pi/2$. Here $z' = 3$, $z'' = -i$ and $Z = 0$ for $t = 2\pi/3$ and $t = 4\pi/3$.

45. Integral of a derivative

Suppose that the function

$$F'(z) = U(x,y) + iV(x,y) \tag{38}$$

is single-valued and has a derivative $F'(z)$ at each point of some domain D, and that this derivative $F'(z)$ is a continuous function in D. Then it follows from Eq. (27) of Section 16 that

$$F'(z) = U_x + iV_x = V_y - iU_y \tag{39}$$

at each point of D.

We may use the derivative $F'(z)$ as the function $f(z)$ of Section 43, if C is any path from z' to z'' in D. This makes

$$F'(z) = f(z) = u + iv. \tag{40}$$

A comparison of Eqs. (39) and (40) shows that

$$u = U_x = V_y \quad \text{and} \quad v = V_x = -U_y. \tag{41}$$

Consequently the integrands in Eq. (29) are

$$\begin{aligned} u\,dx - v\,dy &= U_x\,dx + U_y\,dy = dU \\ v\,dx + u\,dy &= V_x\,dx + V_y\,dy = dV. \end{aligned} \tag{42}$$

Since each of these is the total differential of a single-valued function, we may evaluate the integrals in Eq. (29) as follows:

$$\begin{aligned} \int_C f(z)\,dz &= \int_C dU + i\int_C dV \\ &= \left[U(x,y) + iV(x,y) \right]_{x',y'}^{x'',y''}, \end{aligned} \tag{43}$$

or

$$\int_{z'}^{z''} f(z)\,dz = F(z'') - F(z'). \tag{44}$$

Thus for suitably restricted paths, the integral of a function known to be the derivative of another function can be found by the familiar rules of integral calculus.

Incidentally, Eq. (44) shows that if $f(z) = F'(z)$ in a domain D, where $F(z)$ is single-valued in D, then the integral of $f(z)$ from z' to z'' is the same for any two paths in D joining z' and z''.

46. An inequality for integrals

We shall now obtain a useful upper bound for $|\int_C f(z)\,dz|$. First let C be the smooth curve of length L of Section 42. Note that the length of a chord $|\Delta z_k|$ cannot exceed its arc Δs_k. Hence we have

$$\sum_{k=1}^{n} |\Delta z_k| \leq \sum_{k=1}^{n} \Delta s_k = L. \tag{45}$$

On C, $|f(z)|$ is real and continuous. Hence it is bounded and
$$|f(z)| \leq M, \tag{46}$$
where M is either the maximum of $|f(z)|$ on C, or any larger number. It follows from Eqs. (45) and (46) that
$$\left|\sum_{k=1}^{n} f(\hat{z}_k)\,\Delta z_k\right| \leq \sum_{k=1}^{n} |f(\hat{z}_k)| \cdot |\Delta z_k| \leq M \sum_{k=1}^{n} |\Delta z_k| \leq ML. \tag{47}$$
By taking the limit as in Eq. (23), we may deduce that
$$\left|\int_C f(z)\,dz\right| \leq ML, \tag{48}$$
which is the inequality we were seeking.

Next let C be a contour made up of several smooth pieces, as in Section 44. Then we may apply the preceding result to each piece and so conclude that *the inequality of Eq. (48) holds for any contour C of length L if $|f(z)| \leq M$ on C.*

Exercise 24

From Eq. (44) deduce that for every path from 0 to A,

1. $\int_0^A dz = A$. 2. $\int_0^A z\,dz = \dfrac{A^2}{2}$. 3. $\int_0^A e^z\,dz = e^A - 1$.

Check Eq. (48), with $M = \max |f(z)|$, by verifying that

4. $I = \int_{z'}^{z''} dz = z'' - z'$, $M = 1$, and for a straight path, $L = |z'' - z'|$ so that $ML = L$. Here $|I| = |L| = ML$.

5. $I = \int_0^A z\,dz = A^2/2$, $A \neq 0$. $M = |A|$, and for a straight path, $L = |A|$, so that $ML = |A|^2$. Here $|I| = |A|^2/2 \leq |A|^2 = ML$.

6. $I = \int_{C_1} |z|\,dz$ with C_1 a straight path from $-i$ to i.

$I = i\int_{-1}^{1} |iy|\,dy = i\int_{-1}^{0} (-y)\,dy + i\int_0^1 y\,dy = i$. $M = 1$, $L = 2$. Here $|I| = 1 < 2 = ML$.

7. $I = \int_{C_2} |z|\,dz$ with C_2 the right half of the unit circle from $-i$ to i, $z = e^{it}$ for $-\pi/2 \leq t \leq \pi/2$.

$$I = \int_{-\pi/2}^{\pi/2} ie^{it}\,dt = 2i. \quad M = 1, \quad L = \pi.$$

Here $|I| = 2 < \pi = ML$.

8. For every path from z' to z'', show that for any positive integer n,
$$\int_{z'}^{z''} (z - z_0)^n \, dz = \frac{(z'' - z_0)^{n+1} - (z' - z_0)^{n+1}}{n+1}.$$

9. Assuming that z_0 is *not* on the path from z' to z'', show that for any positive integer $n > 1$,
$$\int_{z'}^{z''} \frac{dz}{(z - z_0)^n} = \frac{(z'' - z_0)^{-n+1} - (z' - z_0)^{-n+1}}{-n+1}.$$

10. Assume that z_0 is *not* on the path from z' to z''. Then there is a determination of $\arg(z - z_0) = \theta(s)$ which is a continuous function of s. Show that
$$\int_{z'}^{z''} \frac{dz}{z - z_0} = \text{Log}\left|\frac{z'' - z_0}{z' - z_0}\right| + i(\theta'' - \theta').$$

11. From Problems 8 to 10, deduce that for C any simple positive circuit about z_0, such as the circle $z = z_0 + ae^{it}$, $0 \leq t \leq 2\pi$,
$$\int_C (z - z_0)^k \, dz = \begin{cases} 2\pi i \text{ for } k = -1. \\ 0 \text{ for } k \text{ integral and } \neq -1. \end{cases}$$

47. Cauchy's integral theorem

Let C be a simple closed contour, so oriented that as we traverse it in the positive direction its interior points are on our left. Let R be the closed region made up of all points interior to or on C. Then Green's theorem asserts that if each of the real functions $P(x,y)$, $Q(x,y)$, $\partial P/\partial y$, $\partial Q/\partial x$ is a continuous function of x and y in R, then

$$\int_C (P \, dx + Q \, dy) = \iint_R \left(\frac{\partial Q}{\partial x} - \frac{\partial P}{\partial y}\right) dx \, dy. \tag{49}$$

Now suppose that the function $f(z)$ of Section 43,
$$f(z) = u(x,y) + iv(x,y), \tag{50}$$
has a derivative $f'(z)$ at each point of R, and that this derivative is a continuous function of z in R. Then as in Section 16,
$$f'(z) = u_x + iv_x = v_y - iu_y. \tag{51}$$

This shows that each of the partial derivatives u_x, v_x, u_y, v_y is a continuous function of x and y in R. Hence we may apply Green's theorem, Eq. (49), to each of the integrals in Eq. (29) and so deduce that

$$\int_C f(z)\,dz = \iint_R (-v_x - u_y)\,dx\,dy + i \iint_R (u_x - v_y)\,dx\,dy. \quad (52)$$

But, because of the Cauchy-Riemann equations $u_x = v_y$, $u_y = -v_x$, each of the integrands on the right is zero for z in R. Hence

$$\int_C f(z)\,dz = 0. \quad (53)$$

This proves one form of *Cauchy's integral theorem*:

If $f(z)$ has a derivative $f'(z)$ which is continuous at each point interior to and on a simple closed contour C, $\int_C f(z)\,dz = 0$.

For R inside a domain D where $f'(z)$ exists, we may prove that Eq. (53) holds without any assumption about the continuity of $f'(z)$. And we shall do this, since this will enable us eventually to deduce the continuity and even the differentiability of $f'(z)$ at all points of an open two-dimensional region, or domain D, from the assumption that $f'(z)$ merely exists throughout D. The stronger theorem is known as the *Cauchy-Goursat integral theorem*, since Goursat was the first to prove it with the weakened hypothesis. We begin by proving it for a triangle.

48. The Cauchy-Goursat integral theorem for a triangle

Let S denote the closed set consisting of the interior and boundary points of a given triangle ABC. We assume that the function $f(z)$ has a derivative $f'(z)$ at each point of S. Then $f(z)$ is necessarily continuous at each point of S. As in Fig. 15, divide triangle ABC into

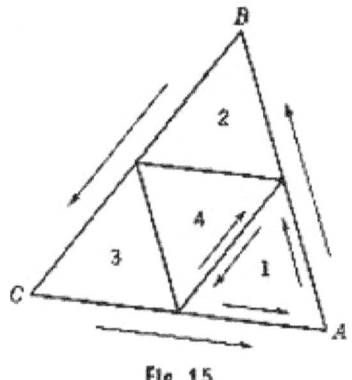

Fig. 15

four triangles of the same shape but with sides half as large by lines through the midpoints of the sides. Let T denote the closed contour bounding triangle ABC, so oriented that as we traverse it in the positive direction the interior points are on our left. And let T_j, $j = 1, 2, 3, 4$, denote similarly oriented closed contours bounding the four smaller triangles. Then we find that

$$\int_T f(z)\, dz = \sum_{j=1}^{4} \int_{T_j} f(z)\, dz, \qquad (54)$$

since T_1, T_2, T_3 each have two sides a part of T, and a third side which is a part of $-T_4$, or T_4 with reversed orientation.

It follows from Eq. (54) that

$$\left| \int_T f(z)\, dz \right| \leq \sum_{j=1}^{4} \left| \int_{T_j} f(z)\, dz \right|. \qquad (55)$$

Let C_1 denote that (or the first) T_j which leads to a term of largest value on the right of Eq. (55). Then we have

$$\left| \int_T f(z)\, dz \right| \leq 4 \left| \int_{C_1} f(z)\, dz \right|, \qquad L_1 = \frac{L}{2}, \qquad (56)$$

where L is the length of T and L_1 is the length of C_1.

We now treat C_1 in the same way we treated the original triangular contour T. This leads to a contour C_2 such that

$$\left| \int_{C_1} f(z)\, dz \right| \leq 4 \left| \int_{C_2} f(z)\, dz \right|, \qquad L_2 = \frac{L_1}{2}, \qquad (57)$$

where L_2 is the length of C_2. Continuing in this way, we obtain a sequence of triangular contours C_n of length L_n such that

$$\left| \int_T f(z)\, dz \right| \leq 4^n \left| \int_{C_n} f(z)\, dz \right|, \qquad L_n = \frac{1}{2^n} L. \qquad (58)$$

Let S_n denote the closed set consisting of C_n and the interior points of the triangle bounded by C_n. Then each point of S_{n+1} is a point of S_n. And the dimensions of S_n tend to zero when n becomes infinite. It follows from the principle of nested closed sets, Problem 1 of Exercise 25, that these sets close down on a single point z_0 which is in each S_n.

Since z_0 is in S_1, z_0 is in S. Thus $f'(z_0)$ exists:

$$\lim_{z \to z_0} \frac{f(z) - f(z_0)}{z - z_0} = f'(z_0). \tag{59}$$

And for any given positive ϵ, there is a δ, such that

$$\left| \frac{f(z) - f(z_0)}{z - z_0} - f'(z_0) \right| \leq \epsilon \quad \text{if} \quad 0 < |z - z_0| < \delta. \tag{60}$$

Consider the function $\omega(z)$ defined by the relations

$$\omega(z) = \frac{f(z) - f(z_0)}{z - z_0} - f'(z_0) \text{ for } z \neq z_0, \quad \omega(z_0) = 0. \tag{61}$$

Since $\omega(z) \to 0$ when $z \to z_0$, by Eq. (59), the definition of $\omega(z_0)$ makes $\omega(z)$ continuous at z_0. Thus for all values of z in S, Eq. (61) implies that $\omega(z)$ is a continuous function and that

$$f(z) = f(z_0) + (z - z_0)f'(z_0) + (z - z_0)\omega(z). \tag{62}$$

We may also conclude from Eqs. (61) and (60) that

$$|\omega(z)| \leq \epsilon \quad \text{if} \quad |z - z_0| < \delta. \tag{63}$$

Let us use Eq. (62) to calculate the integral of $f(z)$ over C_n. The first two terms on the right are the derivative of

$$F(z) = f(z_0)z + \frac{(z - z_0)^2}{2} f'(z_0). \tag{64}$$

This is single-valued for every z. And the path of integration C_n is closed. Thus $z'' = z'$ in Eq. (44) and we have

$$\int_{C_n} [f(z_0) + (z - z_0)f'(z_0)] \, dz = F(z') - F(z') = 0. \tag{65}$$

It follows from Eqs. (62) and (65) that

$$\int_{C_n} f(z) \, dz = \int_{C_n} (z - z_0)\omega(z) \, dz. \tag{66}$$

Now select a particular n so large that $2^n > L/\delta$. Then $L_n = L/2^n < \delta$. And for any point z on C_n, $|z - z_0| < L_n < \delta$, since z_0 is in S_n. Hence for z on C_n, $|\omega(z)| \leq \epsilon$ by Eq. (63), so that $|(z - z_0)\omega(z)| \leq L_n\epsilon$. From this and Eq. (48) we have

$$\left| \int_{C_n} (z - z_0)\omega(z) \, dz \right| \leq (L_n\epsilon)L_n = \epsilon L_n^2. \tag{67}$$

In view of Eq. (66) we may conclude that

$$\left|\int_{C_n} f(z)\, dz\right| \leq \epsilon L_n^2 = \epsilon \left(\frac{L}{2^n}\right)^2 = \frac{1}{4^n} \epsilon L^2. \tag{68}$$

From Eqs. (58) and (68) we may deduce that

$$\left|\int_T f(z)\, dz\right| \leq \epsilon L^2. \tag{69}$$

Since L is fixed, and ϵ is arbitrary, this implies that

$$\left|\int_T f(z)\, dz\right| = 0, \quad \text{and hence} \quad \int_T f(z)\, dz = 0. \tag{70}$$

This proves that if $f(z)$ has a derivative $f'(z)$ at each point interior to and on a triangular contour T, then $\int_T f(z)\, dz = 0$.

49. The Cauchy-Goursat integral theorem

An open two-dimensional region or domain D is said to be *simply connected* if every simple closed polygon consisting entirely of points of D has all its interior points in D. We shall now prove the following form of the Cauchy-Goursat theorem:

Let $f(z)$ have a derivative $f'(z)$ at each point of a simply connected domain D. And let every point of the closed contour C be in D. Then $\int_C f(z)\, dz = 0$.

We first observe that every simple polygon may be decomposed into triangles, by line segments contained in the closed region made up of the polygon and its interior points. The integral of $f(z)$ around the polygon will be the sum of the integrals taken around the component triangles. It will therefore be zero, if the points of the polygon are in D. This result also holds for a polygon which intersects itself, since the integral may be decomposed into a finite number of integrals over simple polygons. Some of these may be degenerate simple polygons consisting of a segment traversed twice in opposite directions.

Consider next any closed contour C which has all its points in D. Then each point s of C is the center of a circle K_s lying in D. But the points of C form a closed and bounded set S. Hence, by the Heine-

Borel theorem of Section 11, we may select a finite number of such circles which cover C. These overlapping circles K together form a closed region R in D which includes C in its interior. Since $f(z)$ is continuous in the closed and bounded region R, by Section 13, it is uniformly continuous in R. That is, for any positive ϵ, there is a δ such that

$$|f(z) - f(z')| \leq \epsilon \text{ for } z, z' \text{ in } R \quad \text{and} \quad |z - z'| < \delta. \qquad (71)$$

Next inscribe a polygon P in the curve C, such that

$$|\Delta z_k| < \delta \quad \text{and} \quad \left| \int_C f(z)\, dz - \Sigma f(z_k)\, \Delta z_k \right| < \epsilon, \qquad (72)$$

where the Δz_k correspond to the sides of the polygon. We may do this, using, among other points of subdivision z_k, the points where the circumferences of the circles K intersect the contour C. This will insure that every point of the polygon P is interior to R.

For the polygon P, we may form a sum approximating the integral of $f(z)$ over the polygon to within ϵ, using points of subdivision z'_j which include the vertices of the polygon. Then

$$\left| \int_P f(z)\, dz - \Sigma f(z'_j)\, \Delta z'_j \right| < \epsilon. \qquad (73)$$

Let z'_{jk} be any point z'_j on the side of the polygon corresponding to Δz_k. Then $|z_k - z'_{jk}| < |\Delta z_k| < \delta$, and by Eq. (71),

$$|f(z_k) - f(z'_{jk})| \leq \epsilon. \qquad (74)$$

Since Δz_k equals the sum of certain $\Delta z'_j$, it follows that

$$|\Sigma f(z_k)\, \Delta z_k - \Sigma f(z'_j)\, \Delta z'_j| \leq \Sigma \epsilon\, |\Delta z'_j| = \epsilon \Sigma |\Delta z_k| \leq \epsilon L, \qquad (75)$$

where L is the length of the closed contour C.

The inequalities (72), (75), and (73) imply that

$$\left| \int_C f(z)\, dz - \int_P f(z)\, dz \right| \leq \epsilon (L + 2). \qquad (76)$$

Since L is fixed and ϵ is arbitrary, it follows that

$$\int_C f(z)\, dz = \int_P f(z)\, dz = 0. \qquad (77)$$

This completes the proof of the Cauchy-Goursat theorem.

Exercise 25

1. The infinite sequence of closed sets S_n is said to be *nested* if each set includes all points of the following set. Let $d_n = \max |z'' - z'|$ for z'', z' in S_n, and let $d_n \to 0$ as $n \to \infty$. Show that there is just one point z_0 which is in every set S_n. *Hint:* Project the sets on the x and y axes, and use the principle for nested closed intervals to obtain x_0 in every X_n, and y_0 in every Y_n. Let $z_0 = x_0 + iy_0$. Select one point in each set, z_n in S_n. From the projections, as $n \to \infty$, $z_n \to z_0$. Since z_m is in S_n for $m > n$, and S_n is closed, the limit point z_0 is in S_n. There could not be two such points z_0 and z_0', since $|z_0' - z_0| \leq d_n$.

Verify that Problem 1 applies to each of the following S_n.

2. S_n: $|z - 2 - 3i| \leq 1/n$, $d_n = 2/n$, $z_0 = 2 + 3i$.
3. S_n: $x \geq 0$, $y \geq 0$, $x + y \leq 2^{-n}$, $d_n = 2^{-n}\sqrt{2}$, $z_0 = 0$.
4. S_n: $n(x^2 + y^2) \leq x$, $d_n = 1/n$, $z_0 = 0$.
5. Show that the conclusion of Problem 1 does not apply to open sets by considering S_n: $n(x^2 + y^2) < x$. *Hint:* If there were a z_0, then $z_0 \neq 0$. This would imply $n < x_0/(x_0^2 + y_0^2)$.
6. That $\int_{C_n} [f(z_0) + (z - z_0)f'(z_0)] \, dz = 0$ was found in Eq. (65). Verify that this follows from the theorem of Section 47, after noting that

$$\frac{d}{dz}[f(z_0) + (z - z_0)f'(z_0)] = f'(z_0)$$

which is constant and hence a continuous function.

7. Prove that if $f(z)$ is *analytic* at each point interior to and on any simple closed contour C, then $\int_C f(z) \, dz = 0$. *Hint:* Apply the reasoning of Section 49, beginning by selecting circles K_s in which the derivative exists.

8. Let C be a *circle*. Let $f(z)$ be continuous in the closed region consisting of the interior and boundary of this circle, and have a derivative $f'(z)$ at each point *interior* to C. Prove that $\int_C f(z) = 0$. *Hint:* Let the circle C be $|z - z_0| = R$. On C let

$$z = z_0 + Re^{i\theta}, \quad dz = Rie^{i\theta} \, d\theta = i(z - z_0) \, d\theta$$

so that

$\int_C f(z) \, dz = \int_0^{2\pi} f(z)(z - z_0)i \, d\theta$. Similarly, on the circle C': $|z' - z_0| = R - \delta$,

$$\int_{C'} f(z')\,dz' = \int_0^{2\pi} f(z')(z'-z_0)i\,d\theta.$$

Now use $\int_{C'} f(z)\,dz = 0$, and the uniform continuity of $f(z)(z-z_0)$, after noting that

$$z(\theta) - z'(\theta) = \delta e^{i\theta}, \quad |z - z'| = \delta.$$

One form of the Cauchy-Goursat theorem states that: *If $f(z)$ has a derivative $f'(z)$ at each point interior to and on a simple closed contour C, then $\int_C f(z)\,dz = 0$.* Prove this for each restricted type of C.

9. C is convex toward all interior points. *Hint:* Use inscribed polygons.

10. C consists of two intersecting straight segments, OA and OB, and a smooth arc AB always convex toward O. *Hint:* Apply the reasoning of Section 49, using circumscribed polygons with vertices z_k''. Take the points of contact as the s_k in Eq. (72). Let the points z_j' include both the z_k and the z_k''. Also take each of $|\Delta z_k|$, $|z_{k+1}'' - z_k|$, and $|z_k - z_{k-1}''|$ less than δ and note that

$$\Delta z_k = (z_{k+1}'' - z_k) + (z_k - z_k'').$$

11. C is smooth and composed of a finite number of pieces, each concave or convex towards some interior point. *Hint:* Use Problems 9 and 10.

50. Multiply connected regions

A domain which is not simply connected is said to be *multiply connected*. A common example is the annular domain between two concentric circles, $r < |z - z_0| < R$, or the limiting case of the deleted neighborhood, $0 < |z - z_0| < R$. More generally, consider a large circle together with a number of smaller circles having no points in common lying inside the large circle. Then the domain made up of those points inside the large circle, but outside of each of the smaller circles, is multiply connected.

Let the function $f(z)$ be single-valued and have a derivative $f'(z)$ at every interior point of a multiply connected domain M. Then the integral around a closed contour in the domain is not necessarily zero. However, if every interior point of the domain bounded by the closed contour belongs to M, the contour is in some simply connected domain to which the theorem of Section 49 applies.

More generally, suppose that we have a number of closed paths C'_j, which together form C', the complete boundary of a multiply connected domain M'. Let C' and M' be interior to M. Then we shall prove that the sum of the integrals of $f(z)$ about the curves C'_j, taken in consistent directions, is zero. We may determine the direction for each C'_j as that in which we traverse C'_j, keeping the adjacent part of M' on our left. With this convention, we have

$$\int_{C'} f(z)\, dz = \sum \int_{C'_j} f(z)\, dz = 0. \tag{78}$$

To prove this, we need merely to separate the multiply connected domain M' into a finite number of simply connected pieces by curves, or crosscuts, lying in M', as in Fig. 16. The integral of $f(z)$ about the

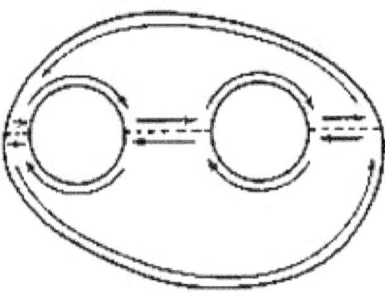

Fig. 16

boundary of each simply connected piece, traversed in a direction which keeps the piece on our left, is zero. In the sum of these integrals, the integration along each crosscut is taken twice, in opposite directions, and thus cancels, since $f(z)$ is single-valued. The remaining boundary of the simply connected pieces makes up the contour C', or the curves C'_j consistently oriented.

51. The Cauchy integral formula

Suppose that $f(z)$ has a derivative at every point of a simply connected domain D. Then if a is any point of D, the function $f(z)/(z-a)$ has a derivative at each point of D except a. Thus we may apply the result of Section 50 to the annular domain M'' inside a simple closed curve C in D, including a as an interior point and outside of a small circle γ with center at a and entirely interior to C. The direction for C is counterclockwise, to have M'' on the left. Hence, if we use C and γ to denote the curves traversed in the positive or counterclockwise direction, the complete boundary of M'' is $C'' = C - \gamma$, and by Eqs. (35) and (78),

$$\int_C \frac{f(z)\,dz}{z-a} - \int_\gamma \frac{f(z)\,dz}{z-a} = 0. \tag{79}$$

Let $\omega(z) = f(z) - f(a)$. Then for any positive ϵ there is a δ such that $|\omega(z)| \leq \epsilon$ if $|z - a| < \delta$, since $f(z)$ is continuous at a. Choose $\rho < \delta$, where ρ is the radius of γ. Then for z on γ, we have

$$f(z) = f(a) + \omega(z), \qquad |\omega(z)| \leq \epsilon. \tag{80}$$

Consequently we have

$$\int_\gamma \frac{f(z)\,dz}{z-a} = \int_\gamma \frac{f(a)\,dz}{z-a} + \int_\gamma \frac{\omega(z)\,dz}{z-a}. \tag{81}$$

For the first integral in the right member, we may set

$$z - a = \rho e^{i\theta}, \quad dz = \rho i e^{i\theta}\,d\theta, \tag{82}$$

$$\int_\gamma \frac{f(a)\,dz}{z-a} = \int_0^{2\pi} f(a)i\,d\theta = 2\pi i f(a). \tag{83}$$

Here θ is a real proper parameter as in Eq. (28).

For the second integral in the right member of Eq. (81), from Eqs. (48) and (80) we have

$$\left|\int_\gamma \frac{\omega(z)\,dz}{z-a}\right| \leq \left(\frac{\epsilon}{\rho}\right) 2\pi\rho = 2\pi\epsilon. \tag{84}$$

The Eqs. (83), (81), and (84) may be combined to give

$$\left|\int_C \frac{f(z)\,dz}{z-a} - 2\pi i f(a)\right| \leq 2\pi\epsilon. \tag{85}$$

Since ϵ is arbitrary, the left member is zero and

$$f(a) = \frac{1}{2\pi i}\int_C \frac{f(z)\,dz}{z-a}. \tag{86}$$

If we denote the complex variable of integration by t, we may replace a by z and write the result in the form

$$f(z) = \frac{1}{2\pi i} \int_C \frac{f(t)\,dt}{t-z}. \tag{87}$$

This is *Cauchy's integral formula*. It expresses the value of $f(z)$ at any point in the domain bounded by C in terms of the integral along C. In the proof, we took C as a single contour bounding a simply connected domain. But the discussion of Section 50 shows that Eq. (87) still holds if the path of integration is C', the properly oriented complete boundary of a multiply connected domain M', provided that z is in M' and that M' is interior to D, a domain in which $f(z)$ is single-valued and has a derivative $f'(z)$.

Let us next assume that $f(z)$ is single-valued and analytic in a domain M', and on its complete boundary C'. Then each point z of C' is the center of some circle K_z in which the derivative $f'(z)$ exists. And the points of C' form a closed and bounded set $\overline{S}$. Hence, by the Heine-Borel theorem of Section 11, we may select a finite number of such circles which cover C'. These overlapping circles K_z, adjoined to M', lead to a domain including M' in which $f(z)$ is single-valued and $f'(z)$ exists. Hence this enlarged domain may be taken as the domain D of the preceding paragraph. This proves the theorem:

Let z be any point of the domain D which may be simply connected or multiply connected. Let C be the properly oriented boundary of D. Then if $f(z)$ is single-valued and analytic in D and on C, Cauchy's integral formula, Eq. (87), is valid.

Corollary 1. Let $g(z)$ be a second function which, like $f(z)$, is single-valued and analytic in D and on C. Then if these functions are equal at all points of the boundary C, $g(t) = f(t)$ for t on C, they must also be equal at all points of the domain D, $g(z) = f(z)$ for z in D.

From $g(t) = f(t)$ for t on C and Eq. (87) we have

$$g(z) = \frac{1}{2\pi i} \int_C \frac{g(t)\,dt}{t-z} = \frac{1}{2\pi i} \int_C \frac{f(t)\,dt}{t-z} = f(z). \tag{88}$$

Corollary 2. Under the conditions of the theorem on D, C, and $f(z)$, the Cauchy-Goursat result, $\int_C f(z)\,dz = 0$, is valid.

Let z_0 be some particular point in D. Then in Eq. (87) with $z = z_0$ we may replace $f(t)$ by $F(t) = (t - z_0)f(t)$. Thus

Sec. 51   INTEGRAL THEOREMS   143

$$\int_C f(t)\,dt = \int_C \frac{F(t)\,dt}{t - z_0} = 2\pi i F(z_0) = 0. \tag{89}$$

Compare Problem 7 of Exercise 25 where a slightly different approach to the conclusion of Corollary 2 was suggested.

Cauchy's integral formula has a number of important consequences, which we shall develop in Chapter 8.

Exercise 26

1. That $\int_C \dfrac{dz}{z - z_0} = 2\pi i$ for C any simple positive circuit about z_0 was found in Problem 11 of Exercise 24. Check this by applying Cauchy's integral formula to $f(z) = 1$.

2. Use the Cauchy-Goursat theorem to show that, if z_0 is exterior to the closed circuit C, then

$$\int_C \frac{dz}{z - z_0} = 0.$$

Let C_1 be the circle $|z| = 1$, C_2 the circle $|z - 3| = 1$, and C_3 the circle $|z| = 9$, each traversed counterclockwise.

3. From Problems 1 and 2 deduce that

$$\int_C \frac{2z - 3}{z^2 - 3z}\,dz = \int_C \frac{dz}{z} + \int_C \frac{dz}{z - 3}$$

equals $2\pi i$ along C_1 or C_2, but equals $4\pi i$ along C_3. Note that along $C_3 - C_1 - C_2$ the integral equals zero, in accordance with Eq. (78).

4. The circuits $-C_1$ and C_3 form the complete boundary of an annulus. Take the straight segment along the real axis joining $z_1 = 1$ and $z_2 = 9$ as a crosscut. Then the closed path consisting of $z_1 z_2$, C_3, $z_2 z_1$, $-C_1$ bounds a simply connected domain D. Evaluate $\int z^{1/2}\,dz$ along this path, assuming that $z^{1/2} = 1$ at z_1 at the start, and varies continuously, so that $z^{1/2} = \sqrt{x}$ along $z_1 z_2$, but $z^{1/2} = -\sqrt{x}$ along $z_2 z_1$.

Ans. $\int_1^9 \sqrt{x}\,dz + \int_0^{2\pi} 3e^{i\theta/2} 9 i e^{i\theta}\,d\theta + \int_9^1 (-\sqrt{x})\,dx +$

$\int_{2\pi}^0 e^{i\theta/2} i e^{i\theta}\,d\theta = 52/3 - 36 + 52/3 + 4/3 = 0.$

5. Check Problem 4 by making another crosscut from -9 to -1, noting that the branch of $z^{1/2}$ which was used in Problem 4 deter-

mines a function which is single-valued and analytic for all interior and boundary points of each half annulus, and applying the Cauchy-Goursat theorem to each half annulus. When adding the two integrals, the values over the new crosscut cancel. But note that here as in Problem 4 the values of the integrals along z_1z_2 and z_2z_1 do *not* cancel. In fact, for the values used

$$\int_{-C_1} z^{1/2}\,dz + \int_{C_1} z^{1/2}\,dz = -\frac{104}{3} \neq 0.$$

This shows that the single-valuedness of $f(z)$ in the multiply connected region M', which was emphasized in Section 50, is a necessary requirement to insure the validity of Eq. (78).

6. Keep the other conditions of Problem 4 unchanged, but let the closed path consist of z_1z_2, C_2 traversed twice, z_2z_1, $-C_1$ traversed twice. Show that the integral over this path is zero by reasoning as in Problem 5. Also show that in this case the integrals along z_1z_2 cancel, so that for the values used here

$$\int_{-2C_1} z^{1/2}\,dz + \int_{2C} z^{1/2}\,dz = 0.$$

These are examples of the integral theorem for the boundary of a simply connected region on the Riemann surface over which the function is single-valued, here the Riemann surface for $z^{1/2}$ described in Section 29.

7. Let $f(z)$ be analytic on a closed contour and in the domain bounded by this positive contour C. Show that

$$\int_C \frac{f(t)\,dt}{t-z} = 2\pi i f(z) \quad \text{or} = 0$$

according as z is interior to C, or exterior to C.

8. Let C_1 and C_2 be two positive closed contours, each of which is exterior to the other. Show that

$$\frac{1}{2\pi i}\left[\int_{C_1}\frac{t^2}{t-z}\,dt + \int_{C_2}\frac{\sin t}{t-z}\,dt\right]$$

$= z^2$ for z inside C_1 and $= \sin z$ for z inside C_2.

9. Let $f(z)$ and $g(1/z)$ be each analytic for $|z| \leq 1$, and let C be the unit circle $|z| = 1$ traversed counterclockwise. Show that

$$I(z) = \frac{1}{2\pi i}\int_C \left[\frac{f(t)}{t-z} + \frac{zg(t)}{zt-t^2}\right]dt$$

$= f(z)$ for $|z| < 1$ and $= g(z)$ for $|z| > 1$. *Hint:* Use Problem 7, and put $t = 1/T$ in the second term of the integrand.

10. Check Problem 9 for $f(z) = 1$, $g(z) = 1$,

$$I(z) = \frac{1}{2\pi i}\int_C \frac{dt}{t} = 1.$$

Also for $f(t) = 0$, $g(t) = 1/t$, and

$$I(z) = \frac{1}{2\pi i}\int_C \left[\frac{1}{t^2} + \frac{1}{zt} - \frac{1}{z(t-z)}\right] dt = \frac{1}{z} - \frac{1}{2\pi i}\frac{1}{z}\int_C \frac{dt}{t-z}.$$

Let $f(z)$ be analytic for $|z| \leq R$, and let C be the circle $|z| = R$ traversed counterclockwise. For z in R, the inverse point in the circle, $Z = R^2/\bar{z}$, is outside the circle.

11. Show that

$$\int_C \frac{f(t)\, dt}{t - Z} = \int_C \frac{\bar{z}f(t)\, dt}{t\bar{z} - R^2} = 0.$$

12. From Problem 11 and Eq. (87) deduce that

$$f(z) = \frac{1}{2\pi i}\int_C \left[\frac{1}{t-z} + \frac{\bar{z}}{R^2 - t\bar{z}}\right] f(t)\, dt$$

$$= \frac{1}{2\pi i}\int_C \frac{(R^2 - z\bar{z})f(t)\, dt}{(t-z)(R^2 - t\bar{z})}.$$

13. With $z = re^{i\phi}$, $t = Re^{i\theta}$, verify that

$$\frac{dt}{(t-z)(R^2 - t\bar{z})} = \frac{dt/t}{(t-z)(\bar{t}-\bar{z})}$$

$$= \frac{i\, d\theta}{R^2 - 2Rr\cos(\theta - \phi) + r^2}.$$

Deduce from Problem 12 that

$$f(z) = \frac{1}{2\pi}\int_0^{2\pi} \frac{(R^2 - r^2)f(Re^{i\theta})\, d\theta}{R^2 - 2Rr\cos(\theta - \phi) + r^2}.$$

Taking real parts gives

$$u(x,y) = u(r\cos\phi, r\sin\phi)$$

$$= \frac{1}{2\pi}\int_0^{2\pi} \frac{(R^2 - r^2)U(\theta)\, d\theta}{R^2 - 2Rr\cos(\theta - \phi) + r^2}.$$

This is *Poisson's integral* which expresses the value of a harmonic function in the interior of a circle of radius R in terms of its values on the boundary of the circle, $U(\theta) = u(R \cos \theta, R \sin \theta)$.

14. Let $f(z)$ be analytic at $z = \infty$ and in the upper half-plane, $\mathrm{Im}\, z = y > 0$. Let the closed contour $C = C_1 + C_2$ with C_1 the semicircle $|z| = S, y \geq 0$ from S to $-S$, and C_2 the real axis from $-S$ to S. Let

$$I = \frac{1}{2\pi i} \int_C \left[\frac{1}{t-z} - \frac{1}{t-\bar{z}} \right] f(t)\, dt,$$

and let I_1, I_2 denote the corresponding integrals over C_1, C_2. Show that for z inside C, $I = f(z)$. Let M_1 be the max $|f(z)|$ on C_1. Show that for $S > 2|z|$, and t on C_1, $|t - z| > S/2$ and

$$|I_1| = \left| \frac{1}{2\pi i} \int_{C_1} \frac{(z-\bar{z})f(t)\, dt}{(t-z)(t-\bar{z})} \right|$$

$$\leq \frac{1}{2\pi} \frac{4}{S^2} 2|y| M_1 (\pi S) = \frac{4|y| M_1}{S},$$

which $\to 0$ when $S \to \infty$ and $M_1 \to f(\infty)$. On C_2, $t = s$, a real parameter, and with $z = x + iy$,

$$I_2 = \frac{1}{\pi} \int_{-S}^{S} \frac{y f(s)\, ds}{(s-x)^2 + y^2}.$$

Deduce that

$$f(z) = \frac{1}{\pi} \int_{-\infty}^{\infty} \frac{y f(s)\, ds}{(s-x)^2 + y^2}.$$

Taking real parts gives

$$u(x,y) = \frac{1}{\pi} \int_{-\infty}^{\infty} \frac{y u(s,0)\, ds}{(s-x)^2 + y^2},$$

Poisson's integral for a half-plane.

15. Verify that the last equation in Problem 14 makes $u(x,y) = 1$ for $u(s,0) = 1$.

16. Let $P(z)$ be a polynomial. Since $P(z) - P(a)$ is zero for $z = a$, it is divisible by $z - a$, and

$$Q(z) = \frac{P(z) - P(a)}{z - a}$$

is a polynomial. Also

$$\int_C \frac{P(z)\, dz}{z - a} = \int_C Q(z)\, dz + \int_C \frac{P(a)\, dz}{z - a}.$$

For C any positive closed circuit about a, $\int_C Q(z)\, dz = 0$ by Cauchy's theorem, and $\int_C dz/(z - a) = 2\pi i$, by Problem 11 of Exercise 24. Deduce that

$$\frac{1}{2\pi i} \int_C \frac{P(z)\, dz}{z - a} = P(a),$$

which gives a simple alternative proof of Cauchy's integral formula for a polynomial.

17. For a circle of radius ρ, let us denote the last term in Eq. (81) by $F(\rho) = \int \omega(z)\, dz/(z - a)$. Although this term seems to depend on ρ, and had to be so considered when deriving Eqs. (84) and (86), show that in fact $F(\rho) = 0$. *Hint:* (a) From the Cauchy-Goursat theorem applied to an annulus, $F(\rho_2) - F(\rho_1) = 0$, so that $F(\rho) = F$, a constant independent of ρ. But by Eq. (84), $|F| \leq 2\pi\epsilon$ for arbitrary ϵ, and $F = 0$. Or (b) From Eq. (86) with $C = \gamma$, first as it stands and then with $f(z) = 1$, we may deduce from Eq. (81) that $2\pi i f(a) = 2\pi i f(a) + F(\rho)$, so that $F(\rho) = 0$.

8

TAYLOR'S EXPANSION

In this chapter we shall show that analytic functions necessarily have Taylor's expansions. Specifically, if $f(z)$ is analytic at a, there is a Taylor's series of positive powers of $(z - a)$ which converges to $f(z)$ for all z in some neighborhood of a. Thus at a the function $f(z)$ must possess derivatives of all orders. This fact is used to establish Morera's theorem, which shows that a certain property of integrals implies analyticity. Several characteristic properties of analytic functions in terms of differentiation, integration, and expansibility in power series are then summarized. Each of these properties could be made the basis of a definition of analyticity.

From the Taylor's expansion we derive integral formulas for $f^{(n)}(a)$. These lead to inequalities for $|f^{(n)}(a)|$. We also establish Bessel's inequality for the coefficients of a Taylor's series, and the maximum modulus principle. We discuss Liouville's theorem, which states that the only function satisfying a certain boundedness condition in the closed plane is a constant, and we use Liouville's theorem to prove the fundamental theorem of algebra.

Taylor's expansions, or functional elements, for the same function but about different points, are interdependent through the process of analytic continuation. This process is described, and used to define a monogenic analytic function. We also define singular points, and prove the monodromy theorem.

52. Integration of uniformly convergent series

In Section 20 we considered an infinite series of functions $\Sigma g_n(z)$. We wrote

$$G(z) = \sum_{n=1}^{\infty} g_n(z), \quad G_N(z) = \sum_{n=1}^{N} g_n(z), \quad R_N(z) = G(z) - G_N(z). \quad (1)$$

We defined the series to be *uniformly convergent* for z in S if for every positive ϵ there exists a single integer $n_1(\epsilon)$ depending only on ϵ and not on z, such that

$$|R_N(z)| < \epsilon \quad \text{for all } z \text{ in } S \quad \text{if } N > n_1(\epsilon). \quad (2)$$

We proved that the sum function $G(z)$ was necessarily continuous in S if each $g_n(z)$ was continuous in S, and the series converged uniformly in S. We shall show that under these conditions the series may be integrated termwise along C, where C is any path of finite length L consisting entirely of points in S. In particular the set S may be the curve C itself.

To prove this, we first observe that in consequence of Eqs. (1), (2), and Eq. (48) of Section 46, we have

$$\left| \int_C G(z) \, dz - \int_C G_N(z) \, dz \right| = \left| \int_C R_N(z) \, dz \right| \leq \epsilon L, \quad (3)$$

if $N > n_1(\epsilon)$. Since L is fixed, it follows that as $N \to \infty$ the left member approaches zero, so that

$$\int_C G(z) \, dz = \lim_{N \to \infty} \int_C G_N(z) \, dz = \sum_{n=1}^{\infty} \int_C g_n(z) \, dz, \quad (4)$$

by Eq. (32) of Section 44. Formally stated, the theorem is:

On C, any contour of finite length, let the infinite series $\Sigma g_n(z)$ converge uniformly to the sum function $G(z)$. Then if each function $g_n(z)$ is continuous on C, Eq. (4) is valid.

53. Taylor's series

Let the function $f(z)$ be single-valued and have a derivative at every point of a domain D. Let a be a point in D. Choose a radius r

follows that the convergence of the series (8) is uniform with respect to t.

Since $|f(t)|$ is real and continuous on C, it is bounded on C. Hence termwise multiplication of the series (8) by $f(t)$ will not disturb the uniform convergence. By the theorem of Section 52, we may integrate the new series termwise, and so deduce from Eq. (5) that

$$f(z) = A_0 + A_1(z - a) + A_2(z - a)^2 + \ldots + A_n(z - a)^n + \ldots, \qquad (10)$$

where

$$A_n = \frac{1}{2\pi i} \int_C \frac{f(t)\,dt}{(t - a)^{n+1}}. \qquad (11)$$

Let $Z = z - a$. Then $z = Z + a$ and $f(z) = f(Z + a)$. Hence

$$F(Z) = f(Z + a) = A_0 + A_1 Z + A_2 Z^2 + \ldots + A_n Z^n + \ldots. \qquad (12)$$

This is a power series in Z of the type discussed in Section 19. Since it converges for all Z with $|Z| < r$, its radius of convergence R is not less than r. Hence for any Z with $|Z| < r$, $|Z| < R$. And by Corollary 3 of Section 21, $F^{(m)}(Z)$ exists and

$$F^{(m)}(Z) = f^{(m)}(Z + a)$$
$$= \sum_{n=m}^{\infty} n(n-1)\ldots(n - m + 1) A_n Z^{n-m}. \qquad (13)$$

Since $Z = z - a$, this is equivalent to

$$f^{(m)}(z) = \sum_{n=m}^{\infty} n(n-1)\ldots(n - m + 1) A_n (z - a)^{n-m}. \qquad (14)$$

As in Corollary 4 of Section 21, Eqs. (10) and (14) imply that

$$f(a) = A_0, \quad f^{(m)}(a) = m! A_m, \quad \text{or} \quad A_n = \frac{1}{n!} f^{(n)}(a). \qquad (15)$$

Thus the series (10) is the Taylor's expansion of $f(z)$ in powers of $(z - a)$. There can be no other series $\Sigma B_n (z - a)^n$ which converges to $f(z)$ for $|z - a| < r_1$ with $r_1 > 0$, since the argument just given would show that

$$B_n = \frac{1}{n!} f^{(n)}(a), \qquad B_n = A_n. \qquad (16)$$

This establishes the following theorem:

Let the function $f(z)$ be single-valued and have a derivative at every point of a domain D. Let a be any point in D. Then there is always one and only one expansion $\Sigma A_n(z-a)^n$ which converges for z in some complete neighborhood of a, $|z-a| < r_1$, $r_1 > 0$, and represents the function $f(z)$ in that neighborhood as in Eq. (10).

Corollary 1. The series is the Taylor's expansion of $f(z)$, with $A_n = (1/n!)f^{(n)}(a)$.

Corollary 2. Each of the higher derivatives of $f(z)$, $f^{(n)}(z)$, exists in D. Hence each of these derivatives is continuous in D, differentiable in D, and has a Taylor's expansion for $|z-a| < r_1$, as in Eq. (14), obtained from Eq. (10) by repeated termwise differentiation.

Corollary 3. The series of Eqs. (10) and (14) converge at least in the largest circle $|z-a| < r_1$ all of whose interior points belong to D. That is, we may take r_1 as the distance from a to the nearest point on the boundary of D.

The exact radius of convergence for all these series $R \geq r_1$ is the least upper bound, which may be ∞, of values R_1 such that within the circle $|z-a| < R_2$, a single-valued differentiable function $g(z)$ can be defined which is equal to $f(z)$ at all points in D with $|z-a| < R_1$.

Let $|z-a| < R_1$ be such a circle. Then we may adjoin its points to D by using $g(z)$ to define the extension of the function $f(z)$ in any added points. This proves that $R \geq R_1$. If R is finite, the series of Eq. (10) defines a suitable function $g(z)$ within the circle $|z-a| < R$, so that R is a possible, and the maximum, R_1. If R is infinite, the series of Eq. (10) defines a suitable function $g(z)$ within any circle $|z-a| < R_1$, and the least upper bound of the R_1 is ∞.

Corollary 4. Let C_1 be the properly oriented complete boundary of D_1, with each point of D_1 and C_1 in D, where $f(z)$ has a derivative. Then if z is in D_1,

$$f^{(n)}(z) = \frac{n!}{2\pi i} \int_{C_1} \frac{f(t)\,dt}{(t-z)^{n+1}} \qquad (17)$$

Let us interpret $f^{(0)}(z)$ to mean $f(z)$, and recall that $0! = 1$. Then for $n = 0$, Eq. (17) becomes Cauchy's integral formula, Eq. (87) of

Section 51. Also, formally, the formula for n could be obtained from that for $n-1$ by differentiation with respect to z under the integral sign. This suggests the possibility of a proof by mathematical induction. See Problems 2 and 3 of Exercise 27. However, we shall give a different proof of Eq. (17) here.

A comparison of Eqs. (11) and (15) shows that

$$f^{(n)}(a) = n!A_n = \frac{n!}{2\pi i} \int_C \frac{f(t)\,dt}{(t-a)^{n+1}}. \tag{18}$$

Let a be any point in D_1. And for the circle $C_1: |z-a| = r$, take the radius r so small that for $|z-a| \leq r$, z is in D_1. Then $-C + C_1$ is the complete boundary of a domain in which $f(t)/(t-a)^{n+1}$ is a single-valued analytic function of t. Hence by the form of the Cauchy-Goursat theorem given in Corollary 2 of Section 51, we have

$$-\int_C \frac{f(t)\,dt}{(t-a)^{n+1}} + \int_{C_1} \frac{f(t)\,dt}{(t-a)^{n+1}} = 0. \tag{19}$$

For any a in D_1, Eqs. (18) and (19) together imply that

$$f^{(n)}(a) = \frac{n!}{2\pi i} \int_{C_1} \frac{f(t)\,dt}{(t-a)^{n+1}}. \tag{20}$$

With a replaced by z, this is Eq. (17) which was to be proved.

Exercise 27

1. Let $G(z,w)$ be a function of two complex variables. Then $G(z,w)$ is said to approach the limit $G(z,w_0)$ uniformly for z in S if, for each positive ϵ, there is a $\delta(\epsilon)$, depending only on ϵ, and not on z, such that $|G(z,w) - G(z,w_0)| < \epsilon$ for all z in S if $|w - w_0| < \delta(\epsilon)$. Let C be any path of finite length L consisting entirely of points of S. And for each w with $|w - w_0| < r$, $r > 0$, let $G(z,w)$ be a continuous function of z. Prove that

$$\lim_{w \to w_0} \int_C G(z,w)\,dz = \int_C G(z,w_0)\,dz.$$

2. Let

$$G(t,\Delta z) = \frac{1}{\Delta z}\left[\frac{f(t)}{t-(z+\Delta z)} - \frac{f(t)}{t-z}\right] \text{ for } \Delta z \neq 0,$$

and let

$$G(t,0) = \frac{\partial}{\partial z}\left[\frac{f(t)}{t-z}\right].$$

Show that

$$|G(t,\Delta z) - G(t,0)| = \left|\frac{f(t)\,\Delta z}{(t - z - \Delta z)(t - z)^2}\right| < \frac{2M|\Delta z|}{|t_1 - z|^3},$$

where $M = \max |f(t)|$ on C_1, when $|\Delta z| < \frac{1}{2}|t_1 - z|$, for t on C_1 and t_1 the point on C_1 nearest to z. Let z be any fixed point in D_1 bounded by C_1 as in Corollary 4. Deduce that as $\Delta z \to 0$, $G(t,\Delta z) \to G(t,0)$ uniformly in t for t on C_1, as defined in Problem 1. Then use Problem 1 to show that

$$\int_{C_1} G(t,\Delta z)\,dt \to \int_{C_1} G(t,0)\,dt.$$

From Cauchy's integral formula,

$$\frac{f(z + \Delta z) - f(z)}{\Delta z} = \frac{1}{2\pi i}\int_{C_1} G(t,\Delta z)\,dt.$$

Let $\Delta z \to 0$, and deduce that $f'(z)$ exists and equals

$$\frac{1}{2\pi i}\int_{C_1}\frac{f(t)\,dt}{(t - z)^2}.$$

This is an alternative proof of Eq. (17) for $n = 1$.

3. Give an alternative proof of Eq. (17) for any n by mathematical induction, using a method similar to that of Problem 2. Here

$$G(t,\Delta z) = \frac{1}{\Delta z}\left[\frac{f(t)}{(t - z - \Delta z)^{n+1}} - \frac{f(t)}{(t - z)^{n+1}}\right]$$

for $\Delta z \neq 0$, and

$$G(t,0) = \frac{\partial}{\partial z}\left[\frac{f(t)}{(t - z)^{n+1}}\right].$$

If t_2 is the point on C_1 farthest from z,

$$|G(t,\Delta z) - G(t,0)| < \frac{(n + 1)(n + 2)(|t_2 - z| + 1)^n M 2^{n+1}|\Delta z|}{|t_1 - z|^{2n+4}},$$

when $|\Delta z| < \dfrac{|t_1 - z|}{2}$, and $|\Delta z| < 1$.

4. For $a \neq 0$, the expansion of $f(z) = 1/z$ may be found from

$$\frac{1}{z} = \frac{1}{a}\left[\frac{1}{1 + (z - a)/a}\right] = \frac{1}{a} - \frac{z - a}{a^2} + \ldots + (-1)^n\frac{(z - a)^n}{a^{n+1}} + \ldots.$$

Verify directly that Eq. (15) holds here by calculating $f^{(n)}(a) =$

$(-1)^n(n!/a^{n+1})$. Also verify directly that Eq. (11) holds here as follows. From Problem 11 of Exercise 24, for C a circle $|z - a| = r < |a|$,
$\int_C \frac{dt}{(t-a)^{n+1}} = 0$ for $n \neq 0$, and $= 2\pi i$ for $n = 0$. Also $\int_C \frac{dt}{t} = 0$
by Cauchy's theorem, since O is outside C. Note that
$$\frac{1}{t(t-a)^{n+1}} = \frac{1}{a}\left[\frac{1}{(t-a)^{n+1}} - \frac{1}{t(t-a)^n}\right].$$
Deduce that
$$A_0 = \frac{1}{2\pi i}\int_C \frac{dt}{t(t-a)} = \frac{1}{a},$$
and if
$$A_n = \frac{1}{2\pi i}\int_C \frac{dt}{t(t-a)^{n+1}} \quad \text{for } n > 0,$$
$A_n = -\frac{1}{a}A_{n-1}$, so that $A_n = (-1)^n \frac{1}{a^{n+1}}$.

5. The function $1/z$ is single-valued and differentiable *within* any circle $|z - a| < R_1$ if $a \neq 0$ and $R_1 \leq |a|$, but *not* at *all* points inside any larger circle $|z - a| < R_2$ with $R_2 > |a|$, since this circle has $z = 0$ as an interior point, and $1/z \to \infty$ as $z \to 0$. Verify directly that for the series of Problem 4, $R = |a|$. This is maximum R_1, in agreement with Corollary 3.

6. Differentiate the series of Problem 4 termwise $(m - 1)$ times to obtain
$$(-1)^{m-1}\frac{(m-1)!}{z^m} = \sum_{n=m-1}^{\infty}(-1)^n \frac{n!(z-a)^{n-m+1}}{(n-m+1)!a^{n+1}}.$$
Deduce that
$$\frac{1}{z^m} = \frac{1}{a^m}\left[1 - m\frac{z-a}{a} + \ldots + (-1)^N\frac{(N+m-1)!z^N}{(m-1)!N!a^N} + \ldots\right].$$
By Corollary 3, $R = |a|$. Verify directly that $R = |a|$. Also check by an argument like that of Problem 5 applied to $f(z) = 1/z^m$. Also verify that the coefficient of z^N is $(1/N!)f^{(N)}(a)$ if $f(z) = 1/z^m$.

7. Put $z - a = Z$ and $a = -b$ in Problem 6 to obtain, for $|Z| < |b|$,
$$\frac{1}{(Z-b)^m} = \left(-\frac{1}{b}\right)^m\left[1 + m\frac{Z}{b} + \ldots \right.$$
$$\left. + \frac{(N+1)\ldots(N+m-1)}{(m-1)!}\cdot\frac{Z^N}{b^N} + \ldots\right] \quad \text{for } m > 1,$$

and
$$\frac{1}{Z-b} = \left(-\frac{1}{b}\right)\left(1 + \frac{Z}{b} + \ldots + \frac{Z^N}{b^N} + \ldots\right).$$

8. Let $E(z)$ be single-valued and satisfy $dE/dz = E$ for $|z| < r$, $r > 0$. Also let $E(0) = 1$. Show that, at least for $|z| < r$, $E(z)$ possesses the Taylor's expansion

$$E(z) = 1 + z + \frac{z^2}{2!} + \ldots + \frac{z^n}{n!} + \ldots.$$

Hence $E(z) = e^z$, since the series, with $R = \infty$, was taken as the definition of e^z in Section 22.

9. Let A be any fixed complex number. Then inside the circle $|z| < 1$, the principal branch of $(1 + z)^A$ is defined by continuity and the condition that its value is unity when $z = 0$. Prove the *binomial theorem:* For $|z| < 1$, this principal branch has the Taylor's expansion,

$$(1+z)^A = 1 + Az + \frac{A(A-1)}{2!}z^2 + \ldots$$
$$+ \frac{A(A-1)(A-2)\ldots(A-n+1)}{n!}z^n + \ldots.$$

10. Prove that for $|z| < 1$, the principal branch of $\log(1+z)$ has the Taylor's expansion

$$\text{Log}(1+z) = z - \frac{z^2}{2} + \ldots + (-1)^{n+1}\frac{z^n}{n} + \ldots.$$

11. Put $Z = z^2$, $b = -1$ in the last equation of Problem 7 to obtain

$$\frac{1}{1+z^2} = 1 - z^2 + \ldots + (-1)^N z^{2N} + \ldots \quad \text{for } |z| < 1.$$

The fact that $1/(1+z^2) \to \infty$ when $z \to i$ or when $z \to -i$ explains why $R = 1$. Without going into the complex plane, it is not easy to see why the corresponding Taylor's expansion for the real function

$$\frac{1}{1+x^2} = 1 - x^2 + \ldots + (-1)^N x^{2N} + \ldots$$

has the radius of convergence unity, since the function $1/(1+x^2)$ has no singularities when we consider only real values of x.

12. Let $S(z)$ and $C(z)$ each be single-valued and satisfy $dS/dz = C$, $dC/dz = -S$ for $|z| < r$, $r > 0$. Also let $S(0) = 0$ and $C(0) = 1$.

Show that, at least for $|z| < r$, these functions have Taylor's expansions,

$$S(z) = z - \frac{z^3}{3!} + \ldots + (-1)^m \frac{z^{2m+1}}{(2m+1)!} + \ldots,$$

and

$$C(z) = 1 - \frac{z^2}{2!} + \ldots + (-1)^m \frac{z^{2m}}{(2m)!} + \ldots.$$

Hence $S(z) = \sin z$ and $C(z) = \cos z$, since these series, each with $R = \infty$, were taken as the definition of $\sin z$ and $\cos z$ in Section 23.

13. Use the result or method of proof of Problem 12 of Exercise 10 to prove the following theorem: Let the two power series $\Sigma A_n(z-a)^n$ and $\Sigma B_n(z-a)^n$ each converge for $|z-a| < r, r > 0$. Then if they have the same sum for any infinite sequence of points $z_n \neq a$ but such that $z_n \to a$ as $n \to \infty$, then $A_n = B_n$.

14. The function $W(z) = 1 + z/2! + z^2/3! + \ldots$ is analytic for all z, and $W(0) = 1$. For $z \neq 0$, $W(z) = (e^z - 1)/z$. Hence the zeros of $W(z)$ nearest the origin are $z = 2\pi i$ and $z = -2\pi i$. Deduce that $1/W(z)$ has an expansion in powers of z, valid for $|z| < 2\pi$. Let

$$\frac{1}{W(z)} = \sum_{n=0}^{\infty} B_n \frac{z^n}{n!}. \quad \text{Then } W(z) \left[\sum_{n=0}^{\infty} B_n \frac{z^n}{n!} \right] = 1.$$

Hence by the Cauchy product rule proved in Problem 5 of Exercise 10, $B_0 = 1$,

$$\sum_{m=0}^{n} \frac{B_{n-m}}{(n-m)!} \frac{1}{(m+1)!} = 0 \quad \text{for} \quad n > 0.$$

Multiplication by $(n+1)!$ makes the coefficients of B_{n-m} binomial coefficients. And the relations for $n = 1$ and $n = 2$ become

$$2B_1 + 1 = 0, \quad 3B_2 + 3B_1 + 1 = 0.$$

For $z \neq 0$,

$$\frac{1}{W(z)} - \frac{1}{W(-z)} = \frac{z}{e^z - 1} + \frac{z}{e^{-z} - 1} = -z.$$

This confirms $B_1 = -\frac{1}{2}$ and shows that $B_{2m+1} = 0$ for $m > 0$. Hence the relation for $n = 4$ simplifies to $5B_4 + 10B_2 + 5B_1 + 1 = 0$. Verify that $B_2 = \frac{1}{6}, B_4 = -\frac{1}{30}, B_6 = \frac{1}{42}$, the *Bernoulli numbers* such that

$$\frac{1}{W(z)} = \frac{z}{e^z - 1} = 1 - \frac{z}{2} + \sum_{m=1}^{\infty} B_{2m} \frac{z^{2m}}{(2m)!}, \text{ for } 0 < |z| < 2\pi.$$

15. $\cot z = i\dfrac{e^{iz} + e^{-iz}}{e^{iz} - e^{-iz}} = i\dfrac{e^{2iz} + 1}{e^{2iz} - 1} = i + \dfrac{2i}{e^{2iz} - 1}$. From Problem 14 deduce that

$$z \cot z = iz + \dfrac{1}{W(2iz)} = 1 + \sum_{m=1}^{\infty} (-1)^m \dfrac{2^{2m}B_{2m}z^{2m}}{(2m)!}, \text{ for } 0 < |z| < \pi.$$

16. $\cot z - \tan z = \dfrac{\cos z}{\sin z} - \dfrac{\sin z}{\cos z} = \dfrac{\cos^2 z - \sin^2 z}{\sin z \cos z} = \dfrac{2\cos 2z}{\sin 2z}.$

From this and Problem 15 deduce that for $|z| < \pi/2$,

$$\tan z = -2\cot 2z + \cot z$$
$$= \sum_{m=1}^{\infty} (-1)^{m-1} \dfrac{(2^{2m} - 1)2^{2m}}{(2m)!} B_{2m} z^{2m-1}.$$

17. The *Legendre polynomial* of the nth degree, $P_n(z)$, may be defined by

$$P_n(z) = \dfrac{1}{2^n n!} \dfrac{d^n}{dz^n} (z^2 - 1)^n.$$

From Eq. (17) deduce that

$$P_n(z) = \dfrac{1}{2\pi i} \int_C \dfrac{(t^2 - 1)^n \, dt}{2^n(t - z)^{n+1}},$$

where C surrounds z. Use this with $t = z + \sqrt{z^2 - 1}e^{i\theta}$ for $z \neq 1$ to deduce Laplace's formula:

$$P_n(z) = \dfrac{1}{2\pi} \int_{-\pi}^{\pi} (z + \sqrt{z^2 - 1} \cos\theta)^n \, d\theta$$
$$= \dfrac{1}{\pi} \int_0^{\pi} (z + \sqrt{z^2 - 1} \cos\theta)^n \, d\theta.$$

For $z = 1$ the above proof fails. But we may note that

$$\dfrac{d^n}{dz^n}\left[(z - 1)^n(z + 1)^n\right]_{z=1} = \left[(z + 1)^n \dfrac{d^n}{dz^n}(z - 1)^n\right]_{z=1} = 2^n n!,$$

so that $P_n(1) = 1$. Hence the result does hold for $z = 1$.

18. Prove that the Bessel's function of order n,

$$J_n(z) = \sum_{m=0}^{\infty} \dfrac{(-1)^m z^{n+2m}}{2^{n+2m} m!(n + m)!}$$

is analytic for all z and is a solution of the differential equation

$$\dfrac{d^2w}{dz^2} + \dfrac{1}{z}\dfrac{dw}{dz} + (1 - \dfrac{n^2}{z^2})w = 0.$$

19. In Eq. (17) put $f(z) = e^{zt/2}$, $n = N + m$, and $z = 0$ to obtain
$$\left(\frac{Z}{2}\right)^{N+m} = \frac{(N+m)!}{2\pi i} \int_C \frac{e^{zt/2} dt}{t^{N+m+1}}$$
if C surrounds the origin. Hence
$$\frac{(-1)^m z^{n+2m}}{2^{n+2m} m!(n+m)!} = \frac{1}{2\pi i} \int_C \frac{1}{m!} \left(-\frac{z}{2t}\right)^m \frac{e^{zt/2}}{t^n} \frac{dt}{t}.$$

Use this with $t = e^{i\theta}$ to prove Bessel's formula for the function of Problem 18:
$$J_n(z) = \frac{1}{2\pi} \int_{-\pi}^{\pi} e^{i(z\sin\theta - n\theta)} d\theta$$
$$= \frac{1}{\pi} \int_0^{\pi} \cos(z\sin\theta - n\theta) d\theta.$$

54. Morera's theorem

In Section 53 we showed that the existence of $f'(z)$ in D implies the existence of each of the higher derivatives $f^{(n)}(z)$ in D. This enables us to prove *Morera's theorem*, which is a converse of the form of the Cauchy-Goursat integral theorem proved in Section 49, namely:

Let $f(z)$ be continuous at each point of a simply connected domain D. Then if $\int_C f(z) dz = 0$ for every closed contour consisting entirely of points of D, the function $f(z)$ has a first derivative $f'(z)$ at each point of D.

To prove this, select any fixed point z_0 in D. Consider the integral of $f(z)$ from z_0 to z, any second point in D, along any path in D. If C_1 and C_2 are any two such paths, $C_1 - C_2$ is a closed contour in D. Hence, by our hypothesis, we must have

$$\int_{C_1} f(z) dz - \int_{C_2} f(z) dz = 0, \quad \int_{C_1} f(z) dz = \int_{C_2} f(z) dz. \quad (21)$$

Thus the integral from z_0 to z is the same for all paths in D, and so may be used to define a single-valued function in D,

$$F(z) = \int_{z_0}^z f(t) dt, \quad \text{over any path in } D. \quad (22)$$

Let $h = \Delta z \neq 0$. Then the difference quotient of $F(z)$ is

$$\frac{\Delta F}{\Delta z} = \frac{F(z+h) - F(z)}{h} = \frac{1}{h} \int_z^{z+h} f(t) dt. \quad (23)$$

Since $f(z)$ is continuous at z, for any positive ϵ there is a δ such that
$$|f(t) - f(z)| < \epsilon, \quad \text{if} \quad |t - z| < \delta. \tag{24}$$
Let us now take $|h| < \delta$, and also so small that all points with $|t - z| < |h|$ lie in D. Then we may put
$$f(t) = f(z) + \omega(t), \quad \text{with} \quad |\omega(t)| < \epsilon \tag{25}$$
for all t on the straight segment joining z and $z + h$. And if we use this as the path of integration in Eq. (23) we have
$$\int_z^{z+h} f(t)\, dt = \int_z^{z+h} [f(z) + \omega(t)]\, dt = hf(z) + \int_z^{z+h} \omega(t)\, dt. \tag{26}$$
From Eq. (48) of Section 46, and Eq. (25), it follows that
$$\left| \int_z^{z+h} \omega(t)\, dt \right| \leq \epsilon |h|. \tag{27}$$
We may conclude from Eqs. (23), (26), and (27) that
$$\left| \frac{\Delta F}{\Delta z} - f(z) \right| \leq \epsilon \tag{28}$$
for $|h| = |\Delta z|$ sufficiently small. It follows from this that
$$\lim_{\Delta z \to 0} \frac{\Delta F}{\Delta z} = f(z). \tag{29}$$

This proves that $F(z)$ has a derivative $F'(z)$ at z, any point in D. Hence, by Corollary 2 of Section 53, each of the higher derivatives of $F(z)$ exists in D. In particular the second derivative $F''(z)$ exists. But since $F'(z) = f(z)$ by Eq. (29), it follows that $F''(z) = f'(z)$. Thus $f'(z)$ exists, as we set out to prove.

55. Analytic functions

According to the definition given in Section 14, a single-valued function $f(z)$ is analytic at every point of an open two-dimensional region, or domain D, if $f'(z)$ exists throughout D. And a function is analytic at a point z if z is in, and hence an interior point of, some domain D throughout which $f'(z)$ exists.

By Section 54, $f(z)$ is analytic in D if it is single-valued and continuous in D, and may be integrated in D in the sense that its integral around every closed contour in a simply connected part of D is zero, or that the integral between any two points is independent of the path, for all choices of path lying in a simply connected part of D.

Similarly, any function which is analytic in D may be integrated in this sense, by the Cauchy-Goursat theorem of Section 49.

By the discussion of Eqs. (10) to (14), or of Eq. (18) in Section 21, it was shown that a function represented by a power series is analytic at all points inside of its circle of convergence. By the theorem of Section 53, a function analytic at a point may be represented by a power series expansion about this point.

By Corollary 2 of Section 53, a function analytic at a point and hence analytic in some D including this point has a first derivative $f'(z)$ which is continuous throughout D. Hence by Section 16, if $f(z) = u + iv$, then u and v will each have first partial derivatives which are continuous and satisfy the Cauchy-Riemann differential equations throughout D. And conversely, if u and v satisfy these two conditions, $f(z) = u + iv$ is analytic throughout D.

We thus have necessary and sufficient conditions for analyticity, or alternative definitions of analyticity, in terms of derivatives, integrals, power series, or the Cauchy-Riemann differential equations. Each of these is a convenient method of proving analyticity under certain conditions.

A function, proved to be analytic in D by any one of these tests, has all the properties just mentioned. Moreover, by Corollary 2 of Section 53, such a function has derivatives of all orders each of which, like itself, is continuous and in fact analytic in D. Also, by the reasoning of Section 54, it follows that with a fixed lower limit the integral of a function over any path lying in a simply connected part of D is a single-valued analytic function of its upper limit in this part.

56. Inequalities

Let the Taylor's expansion of Eq. (10),

$$f(z) = \sum_{n=0}^{\infty} A_n (z - a)^n, \qquad (30)$$

have a radius of convergence R, where $R > 0$ and may be ∞. Let

C_r denote the circle $|z - a| = r$ traversed counterclockwise, where $r < R$. Then from Eq. (11) we have

$$A_n = \frac{1}{2\pi i} \int_{C_r} \frac{f(t)\,dt}{(t-a)^{n+1}}. \tag{31}$$

Let M_r denote the maximum value of $|f(z)|$ on C_r. Then from Eq. (48) of Section 46 we find that

$$|A_n| = \left| \frac{1}{2\pi i} \int_{C_r} \frac{f(t)\,dt}{(t-a)^{n+1}} \right| \leq \frac{1}{2\pi} \left(\frac{M_r}{r^{n+1}} \right) (2\pi r) = \frac{M_r}{r^n}. \tag{32}$$

This and Eq. (15) establish *Cauchy's inequalities*:

$$|f(a)| = |A_0| \leq M_r, \quad \frac{1}{n!} |f^{(n)}(a)| = |A_n| \leq \frac{M_r}{r^n}. \tag{33}$$

For z on C_r, the series in Eq. (30) converges absolutely. And so does the series in the relation

$$\overline{f(z)} = \sum_{n=0}^{\infty} \overline{A_n}(\bar{z} - \bar{a})^n, \tag{34}$$

since conjugate numbers have the same absolute values. Hence

$$|f(z)|^2 = f(z)\overline{f(z)}$$
$$= \lim_{N \to \infty} \sum_{n=0}^{N} \sum_{m=0}^{N} A_n \overline{A_m}(z-a)^n(\bar{z}-\bar{a})^m. \tag{35}$$

Let $r' = \frac{1}{2}(r + R)$ if R is finite, and $r' = r + 1$ if $R = \infty$. Then $r < r' < R$, so that by Eq. (33) $|A_n r'^n| \leq M_{r'}$. And

$$c = \frac{r}{r'} < 1, \quad |A_n r^n| = |A_n r'^n c^n| \leq M_{r'} c^n. \tag{36}$$

For z on C_r, $|z - a| = |\bar{z} - \bar{a}| = r$, and

$$|A_n \overline{A_m}(z-a)^n(\bar{z}-\bar{a})^m| = |A_n r^n A_m r^m| \leq M_{r'}^2 c^{n+m}. \tag{37}$$

But for $c < 1$, the double series of positive constants

$$\sum_{n=0}^{\infty} \sum_{m=0}^{\infty} c^{n+m} = \lim_{N \to \infty} \sum_{n=0}^{N} \sum_{m=0}^{N} c^{n+m}$$
$$= \frac{1}{(1-c)^2} \tag{38}$$

converges. Hence the limit on the right in Eq. (35) is approached uniformly for z on C_r by the Weierstrass M test. For z on C_r, we may put $z = a + re^{i\theta}$ and write

$$|f(a + re^{i\theta})|^2 = \lim_{N \to \infty} \sum_{n=0}^{N} \sum_{m=0}^{N} A_n \overline{A_m} r^{n+m} e^{i(n-m)\theta}. \tag{39}$$

Because of the uniformity in θ, as in Section 52, we may integrate with respect to θ termwise from 0 to 2π. We note that

$$\int_0^{2\pi} e^{i(n-m)\theta} d\theta = 0 \text{ for } n \neq m, \text{ but } = 2\pi \text{ for } n = m. \tag{40}$$

Thus the result of the integration is

$$\int_0^{2\pi} |f(a + re^{i\theta})|^2 d\theta = 2\pi \sum_{n=0}^{\infty} A_n \overline{A_n} r^{2n}$$

$$= 2\pi \sum_{n=0}^{\infty} |A_n|^2 r^{2n}. \tag{41}$$

This, or the equivalent form,

$$\sum_{n=0}^{\infty} |A_n|^2 r^{2n} = \frac{1}{2\pi} \int_0^{2\pi} |f(a + re^{i\theta})|^2 d\theta \tag{42}$$

is known as *Parseval's identity*. We recall that

$$M_r = \max |f(z)| \quad \text{on} \quad C_r = \max |f(a + re^{i\theta})|. \tag{43}$$

It follows that the right member of Eq. (42) cannot exceed M_r^2, and

$$\sum_{n=0}^{\infty} |A_n|^2 r^{2n} \leq M_r^2. \tag{44}$$

This is *Bessel's inequality*. Since each term on the left is positive, we may replace the infinite sum by any partial sum and in particular by just one term without destroying the inequality. This gives a second proof of Eq. (33), which shows when the equality can hold in that equation. Suppose that for any $r < R$ the equality holds in Eq. (33) for some one value of n, say $n = m$. Then Eq. (44) shows that $A_n = 0$ for $n \neq m$. Hence by Eq. (30), $f(z) = A_m(z - a)^m$. For $m = 0$ the result is *if $|f(z)| = M_r$ for any $r < R$, then we must have $f(z) = A_0 = f(a)$, a constant.*

57. The maximum principle

We have just seen that the equality $|f(a)| = M_r$ for any $r < R$ implies that the function $f(z)$ of Eq. (30) is a constant. Thus when the function $f(z)$ is not a constant, the first relation of Eq. (33) must be an inequality $|f(a)| < M_r$ for every $r < R$. Let N denote any neighborhood of $z = a$. Choose r so small that the circle $C_r: |z - a| = r$ lies in N. Then $|f(z)|$ is a real and continuous function on the closed set C_r and so assumes its maximum M_r at some point of C_r, say z_1. Thus we have $|f(z_1)| = M_r$ and $|f(a)| < |f(z_1)|$. That is, in N there is a point z_1 at which $|f(z)|$ exceeds $|f(a)|$. This proves the preliminary modulus principle:

If $f(z)$ is analytic at $z = a$, and is not a constant, every neighborhood of a contains points z_1 at which $|f(z_1)|$ exceeds $|f(a)|$.

This preliminary result has as an immediate consequence the following *maximum principle*, or *maximum modulus theorem*:

Let C be the boundary of the domain D. Let $f(z)$ be analytic, but not a constant, in D and continuous on $D + C$. Then $|f(z)|$ cannot assume its maximum value in the closed set $D + C$ at an interior point, or point of D.

Corollary. The maximum value of $|f(z)|$ in $D + C$ is attained at some point on the boundary C.

The conditions make $|f(z)|$ a continuous function on the closed set $D + C$. Hence the maximum is attained at some point of $D + C$. By the theorem, it is not in D. Hence it must be on the boundary C.

We state as another form of the maximum modulus theorem:

Let C be the boundary of the domain D. Let $f(z)$ be analytic in D and continuous on $D + C$. If $|f(z)| \leq M$ on C, then $|f(z)| < M$ at all points of D unless $f(z)$ is a constant, and $|f(z)| = M$.

Let M_1 be the maximum value of $|f(z)|$ in $D + C$, attained at z_1. If $f(z)$ is constant, $|f(z)| = M_1$ on $D + C$. Either $M > M_1$ and $|f(z)| < M$, or $M = M_1$ and $|f(z)| = M$, the exceptional case. Assume that $f(z)$ is not constant. Then by the corollary, z_1 is on C, and for z in D, $|f(z)| < |f(z_1)| \leq M$.

If $f(z) \neq 0$ in D, we also have a *minimum modulus theorem*:

Let C be the boundary of the domain D. Let $f(z)$ be analytic, but not a constant, in D and continuous on $D + C$. Let $f(z) \neq 0$ for z in D. Then $|f(z)|$ cannot assume its minimum value in the closed set $D + C$ at an interior point, or point of D. But the minimum value of $|f(z)|$ in $D + C$ is attained at some point of the boundary C.

If for some z_1 on C, $f(z_1) = 0$, then $|f(z)| > 0$ for z in D, and the minimum value of $|f(z)|$ for z in $D + C$ is attained at z_1 since $|f(z_1)| = 0$. Next assume that $f(z) \neq 0$ on C, and therefore in $D + C$. Then $1/f(z)$ is analytic in D and continuous on $D + C$, and $|1/f(z)|$ is a maximum when $|f(z)|$ is a minimum. Thus the result follows from the maximum principle applied to $1/f(z)$.

Both the maximum and the minimum principle apply to harmonic functions, and we have the following theorem:

Let C be the boundary of the domain D. Let $u(x,y)$ be harmonic, but not a constant, in D and continuous on $D + C$. Then $u(x,y)$ cannot assume its maximum or its minimum value in the closed set $D + C$ at an interior point, or point of D. But these will be attained at some points of the boundary C.

To prove this, let a be any point of D. Choose r_1 sufficiently small so that all points of $N_1: |z - a| \leq r_1$ belong to D. Then by Problem 1 of Exercise 8 there exists a conjugate function $v(x,y)$ such that $u + iv$ is an analytic and single-valued function of $z = x + iy$ in N_1. Hence $f(z) = e^{u+iv}$ is analytic in N_1. Also $|f(z)| = e^u \neq 0$ for z in N_1. And $f(z)$ is not a constant, since u is not a constant. Hence the maximum and minimum modulus principle applies to $f(z)$ in N_1, so that e^u cannot assume its maximum or minimum in N_1 for $x + iy = a$. Since e^u is a monotonic function of u, the same is true of u. Since u cannot assume its maximum or minimum in N_1 at a, it cannot assume its maximum or minimum for $D + C$ at a, any point in D. Hence it must assume these values at points on C.

58. Liouville's theorem

As we indicated in Section 22, a function which is single-valued and analytic at all finite points of the plane is called an entire function. Concerning such functions we shall now prove *Liouville's theorem:*

Let $f(z)$ be an entire function. If $|f(z)|$ is uniformly bounded, $|f(z)| < M$, for all finite values of z, then $f(z)$ must be a constant.

With $a = 0$, the Taylor's expansion of Eq. (10) is

$$f(z) = A_0 + A_1 z + A_2 z^2 + \ldots + A_n z^n + \ldots \qquad (45)$$

Since $f(z)$ is an entire function, we may take D as the open complex plane. Hence $R = \infty$ by Corollary 3 of Section 53. And Eq. (33) is valid for any value of r. But $M_r < M$, so that

$$|A_n| \leq \frac{M_r}{r^n} < \frac{M}{r^n}, \qquad n \geq 1. \qquad (46)$$

For r large, $1/r$ and $1/r^n$ are small. Since r is arbitrary, it follows from Eq. (46) that $A_n = 0$ for $n \geq 1$. Hence

$$f(z) = A_0, \qquad (47)$$

a constant, as was to be proved.

We may use Liouville's theorem to give a short proof of the *fundamental theorem of algebra:*

Let $P(z)$ be a polynomial in z of degree one or greater,

$$P(z) = a_0 + a_1 z + a_2 z^2 + \ldots + a_n z^n, \qquad n \geq 1, a_n \neq 0. \qquad (48)$$

Then the equation $P(z) = 0$ has at least one root.

Suppose that $P(z)$ is not zero for any value of z. Then the function $f(z) = 1/P(z)$ is an entire function. Moreover,

$$\lim_{z \to \infty} \frac{P(z)}{a_n z^n} = 1, \quad \lim_{z \to \infty} f(z) = \lim_{z \to \infty} \frac{1}{a_n z^n} = 0. \qquad (49)$$

Hence there is an r_1 such that $|f(z)| < 1$ for $|z| > r_1$. In the closed region $|z| \leq r_1$, $|f(z)|$ is continuous and therefore has a maximum M_1. Then $|f(z)| < M_1 + 1$ for all z, so that by Liouville's theorem $f(z)$ is a constant, $f(z) = f(0)$. But

$$f(z) = f(0) = \frac{1}{P(0)} \neq 0 \quad \text{and} \quad \lim_{z \to \infty} f(z) = f(0) \neq 0. \qquad (50)$$

This contradicts Eq. (49) and so proves the assumption that $P(z) = 0$ had no roots to be untenable.

As a generalization of Liouville's result, we have the theorem:

Let $f(z)$ be an entire function. If $|f(z)| < K|z|^p$, where K and p are positive constants, for all sufficiently large finite values of z, $|z| > r_1$, then $f(z)$ must be a polynomial of degree not exceeding p.

We again use Eq. (45) with $R = \infty$. And we note that Eq. (33) is valid for any value of r. But here for $r > r_1$, $M_r < Kr^p$ and

$$|A_n| \leq \frac{M_r}{r^n} < \frac{K}{r^{n-p}}, \quad n > p, r > r_1. \tag{51}$$

Since r may be arbitrarily large, $1/r$ and $1/r^{n-p}$ may be arbitrarily small. It follows that $A_n = 0$ for $n > p$. Hence

$$f(z) = A_0 + A_1 z + A_2 z^2 + \ldots + A_N z^N, \quad N \leq p, \tag{52}$$

a polynomial of degree not exceeding p, as was to be proved.

Exercise 28

1. In Section 45, we derived the equation

$$\int_{z'}^{z''} f(z)\, dz = F(z'') - F(z')$$

for z', z'', and the path joining them in D, on the assumption that an indefinite integral $F(z)$, single-valued, analytic, and with $F'(z) = f(z)$ existed in D. Verify that when D is simply connected, Eq. (22) with any choice of z_0 in D may be used to construct an indefinite integral. Also that the indefinite integrals so obtained from two different points, $F_1(z)$ from $z_0 = z_1$ and $F_2(z)$ from $z_0 = z_2$ differ by a constant,

$$F_2(z) - F_1(z) = \int_{z_2}^{z_1} f(t)\, dt.$$

2. Use Theorem 2 of Section 16 to show that *any* two indefinite integrals as defined for D in Section 45 for a given function $f(z)$ can differ at most by a constant. This checks the last result of Problem 1.

Verify each of the following instances of Eq. (22), with D the whole open complex plane.

3. $\int_0^z \cos t\, dt = \sin z.$ 4. $\int_{\pi/2}^z \sin t\, dt = -\cos z.$

5. $\int_0^z \cosh t\, dt = \sinh z.$ 6. $\int_{i\pi/2}^z \sinh t\, dt = \cosh z.$

7. Verify that $\int_1^z \frac{dt}{t} = \text{Log } z$ for any path in the cut plane formed by omitting the negative real axis.

8. Verify that $\int_{-1}^z \frac{dt}{t} = \text{Log } (-z)$ for any path in the cut plane formed by omitting the positive real axis.

9. For $f(z) = e^z$ and $a = 0$, verify that $M_r = e^r$, $A_n = 1/n!$ so that Eq. (33) becomes $1/n! < e^r/r^n$, and Eq. (42) becomes

$$\frac{1}{2\pi}\int_0^{2\pi} e^{2r\cos\theta}\, d\theta = \sum_{n=0}^{\infty} \frac{r^{2n}}{(n!)^2}.$$

10. Check the last relation of Problem 9 by deducing from Problems 18 and 19 of Exercise 27 that

$$\sum_{m=0}^{\infty} \frac{r^{2m}}{(m!)^2} = J_0(-2ir) = \frac{1}{2\pi}\int_{-\pi}^{\pi} e^{2r\sin\theta}\, d\theta$$
$$= \frac{1}{2\pi}\int_0^{2\pi} e^{2r\cos\theta}\, d\theta.$$

11. For $f(z) = 1/(1-z)$, $a = 0$, and $r < 1$, verify that $M_r = 1/(1-r)$, $A_n = 1$, so that Eq. (33) becomes $1 < 1/r^n(1-r)$ and Eq. (42) becomes

$$\frac{1}{2\pi}\int_0^{2\pi} \frac{d\theta}{1 - 2r\cos\theta + r^2} = \sum_{n=1}^{\infty} r^{2n} = \frac{1}{1-r^2}.$$

12. Check the last relation of Problem 11 by putting $\phi = 0$, $U(\theta) = 1$ in Poisson's integral found in Problem 13 of Exercise 26 to obtain

$$\frac{1}{2\pi}\int_0^{2\pi} \frac{(1-r^2)\, d\theta}{1 - 2r\cos\theta + r^2} = u(r,\theta) = 1.$$

13. For $f(z) = \frac{1}{(1-z)^2}$, $a = 0$ and $r < 1$, verify that $M_r = \frac{1}{(1-r)^2}$, $A_n = n + 1$, so that Eq. (33) becomes $n + 1 < \frac{1}{r^n(1-r)^2}$ and Eq. (42) becomes

$$\frac{1}{2\pi}\int_0^{2\pi} \frac{d\theta}{(1 - 2r\cos\theta + r^2)^2} = \sum_{n=0}^{\infty}(n+1)^2 r^{2n} = \frac{1+r^2}{(1-r^2)^3}.$$

14. Let $f(z) = 1 + z + z^2 + \ldots + z^{n-1}$, $a = 0$ and $r = 1$ in Eq. (33). Thus deduce that

$$\frac{1}{2\pi}\int_0^{2\pi}\left[\frac{\sin n\theta/2}{\sin \theta/2}\right]^2 d\theta = n.$$

Hint: Note that for $z \neq 1$, $f(z) = (z^n - 1)/(z - 1)$, and if $z = e^{i\theta}$, $z^n - 1 = 2e^{in\theta/2}\sin(n\theta/2)$, $z - 1 = 2e^{i\theta}\sin(\theta/2)$.

15. Let C be the boundary of D. Let $f(z)$ be analytic, but not a constant, in D and continuous on $D + C$. Prove that if $|f(z)| = K$, a constant, on C, there must be at least one point z_1 in D at which $f(z_1) = 0$.

16. Let C be the boundary of D. Let $u(x,y)$ be harmonic in D, and continuous on $D + C$. Prove that if $u(x,y) = K$, a constant, on C, then $u(x,y)$ must be constant in $D + C$.

17. Let $\overline{PP_j}$ denote the distance from the variable point P to the fixed point P_j. For P any point z of the closed region $|z| \leq r$, prove that the maximum value of the product $\overline{PP_1} \cdot \overline{PP_2} \cdot \overline{PP_3}$ is attained at some point on the circle $|z| = r$. *Hint:* The product is equal to $|(z - z_1)(z - z_2)(z - z_3)|$.

18. In Problem 17, prove that the minimum value of the product is also attained at some point on the circle $|z| = r$ if each of the points P_1, P_2, P_3 is outside the circle $|z| = r$. If any one of these points, say P_1, is inside the circle, the minimum zero is attained at P_1.

19. Prove the analogue of Liouville's theorem for harmonic functions: *Let the function $u(x,y)$ be harmonic at all finite points of the plane. If $u(x,y)$ is uniformly bounded, $u(x,y) < K$, then $u(x,y)$ must be a constant.* *Hint:* By Problem 1 of Exercise 8, there exists a conjugate function $v(x,y)$ such that $u + iv$ is analytic in the open plane. Then $f(z) = e^{u+iv}$ is an entire function, with $|f(z)| = e^u < e^K$. Hence by Liouville's theorem, e^{u+iv} is constant, which makes u a constant.

20. Prove the fundamental theorem of algebra by deriving Eq. (49) as in the text, applying the maximum modulus principle for $f(z)$ to the closed circular region $|z| \leq N$, and letting $N = 1, 2, 3, \ldots$. *Hint:* This leads to an increasing sequence of positive maxima, M_N, which is absurd, since $\lim\limits_{N \to \infty} M_N = 0$ by Eq. (49).

21. Prove the fundamental theorem of algebra by applying the minimum modulus principle for $P(z)$ to the closed circular region $|z| \leq N$, and letting $N = 1, 2, 3, \ldots$. *Hint:* If $P(z) \neq 0$, the minimum m_N on $|z| = N$ is one of a decreasing sequence of positive minima, m_N. This is absurd since

$$\lim_{N \to \infty} m_N = \lim_{|z| \to \infty} P(z) = \infty.$$

22. Let $P(z)$ be any polynomial of degree n, with $n \geq 1$, as in Eq. (48). Prove that

$$P(z) = a_n \prod_{s=1}^{m} (z - b_s)^{q_s},$$

where the b_s are distinct and $\sum_{s=1}^{m} q_s = n$. Also show that the decomposition is unique except for the order of the b_s. *Hint:* For $n = 1$, $a_0 + a_1 z = a_1(z - b_1)$ if $b_1 = -a_0/a_1$. For $n > 1$, for some b, $P(b) = 0$ by the fundamental theorem of algebra. Since $P^{(n)}(z) = n!a_n$, all higher derivatives are zero, and the Taylor's expansion in powers of $(z - b)$ has the form

$$P(z) = A_1(z - b) + \ldots + A_{n-1}(z - b)^{n-1} + a_n(z - b)^n.$$

Hence $P(z) = a_n(z - b)Q(z)$, where $Q(z)$ is a polynomial of degree $(n - 1)$ with z^{n-1} as its term of highest degree. Now use mathematical induction. The b_s are uniquely determined as the distinct values at which $P(b_s) = 0$. And for each b_s, q_s is determined as the largest positive integer q for which

$$\lim_{z \to b_s} \frac{P(z)}{(z - b_s)^q} \quad \text{is finite.}$$

23. Let $f(z)$ be an entire function. Prove that if $|f(z)| < K \text{Log} |z|$, where K is a constant, for all sufficiently large values of z, $|z| > r_1$, then $f(z)$ must be a constant.

59. Analytic continuation

Suppose that a function $f(z)$ is single-valued and analytic at all points of some simply connected domain D. Then, by Section 53, with each point a of D there is associated a Taylor's expansion of $f(z)$ in powers of $(z - a)$. The coefficients may be found from the successive derivatives of $f(z)$ at a. And the series has a positive radius of convergence. We call this power series expansion about the point a the *function element*, or simply the *element* of $f(z)$ at a.

We shall now show that the element at any one point z_0 of D determines the element at every other point z of D. Let z be any particular point of D distinct from z_0. Then we may join z_0 to z by a curve lying in D. Since this curve is composed of interior points and with its end points forms a closed set, there is a distance d such that a circle of radius d, with center at any point of the curve, lies in D. Now take a sequence of points $z_0, z_1, z_2, \ldots, z_n = z$ on the curve such that $|z_{k+1} - z_k| < d$. Then the Taylor's expansion at z_0 is valid at z_1. Hence it determines the value of $f(z)$ as well as all the derivatives of $f(z)$, and so the Taylor's expansion or element at z_1. Similarly each element at z_k determines the element at z_{k+1} and so finally the value of the element at z.

This process of finding the element at the point z from that at z_0 is known as *analytic continuation*.

The possibility of analytic continuation proves that any conditions which determine the value of the function and all its derivatives at z_0, for example, the value on any arc through z_0, determine the value of the function in D. In particular if the function is zero along any arc and analytic in D, it must be zero throughout D.

This establishes the "principle of permanence of form," which in the first instance states that if $f(z) = g(z)$ along any arc inside a simply connected domain D in which both functions are single-valued and analytic, the equation holds for all values in D. Under the assumptions, $f(z) - g(z)$ is zero on the arc and therefore throughout D. The principle may be extended to identities or differential equations involving several functions. In particular, for algebraic identities, we formulate the theorem:

Let each of the functions $f_k(z)$ be single-valued and analytic in a simply connected domain D. And let $P(f_k)$ be a polynomial in the n variables $f_1, f_2, \ldots, f_n$. Then if $P[f_k(z)] = 0$ along any arc in D, this relation holds for all values in D.

The definitions in Chapter 4 led to values of e^z, $\sin z$, and $\cos z$ for complex values which yielded functions analytic throughout the open plane, and agreed with the definitions of e^x, $\sin x$, and $\cos x$ for z on the real axis. It follows from our first instance of the principle that no values differing from these could yield functions with these two properties. And the theorem on algebraic identities explains why all the familiar relations of the functions originally established only for real values continued to hold throughout the complex plane.

60. Definition by continuation

In the preceding section we began with a global definition of $f(z)$ which provided its values at all points under consideration. We then showed that the function elements at different points were strongly interdependent. We now wish to use this dependence to define the function globally from the knowledge of a single function element.

If we start with the element at a point and a curve through the point, it may be possible to obtain a function which is single-valued and analytic at all points of the curve by the process of analytic continuation of Section 59. Now consider a domain D consisting of those points inside some circle with center at A where $z = a$ and not lying on a particular radius AB. Suppose that, starting with an element at some point of D, the process of analytic continuation leads to a function single-valued in D, but to two different elements for each inner point of the radius AB, depending on the side of approach, if we attempt to extend the continuation across AB. Then the elements in D determine one branch of a multiple-valued function. And a is called a *branch point* for this branch.

In place of defining a branch by means of a radius AB, we may define a branch by means of some other contour which joins A to B, on the circle, and so prevents us from drawing a closed curve surrounding A in D. We modify the definition of *branch point* accordingly. Such contours AB are known as *branch cuts*.

Let the domain D' consist of all the points outside a circle about the origin, with the exception of those points on some contour extending from a point of this circle to infinity. And let D denote the domain in the Z-plane into which D' is mapped by the transformation $Z = 1/z$. Then if a branch of the function $f(z)$ defined in D' is such that the corresponding branch of $f(1/Z)$ in D has a branch point at the origin, we say that $f(z)$ has a *branch point* at infinity.

The definitions of branch point and branch cut given here are consistent with those given in Section 29 where these terms were illustrated for some simple globally defined functions such as $\sqrt{z}$ and $\log z$.

Let us start with a single function element, and imagine *all* the other elements which can be obtained from it by analytic continuation. The totality of these elements make up a *monogenic analytic function* as defined by Weierstrass. We sometimes use the simpler term *analytic function*. Thus two different analytic expressions represent the same analytic function in this sense if and only if the values of one can be obtained by analytic continuation from some function element representing values of the other. The complete monogenic analytic function need not be a single-valued function.

Consider an arc joining z_1 to z_2. Suppose that it is possible, starting with an element at z_1, to obtain an element for each point of the arc between z_1 and z_2 but not for z_2 itself by analytic continuation. And let these elements all belong to a single-valued branch of the function. Then z_2 is called a *singularity* or *singular point* for this branch. The singularity may be a branch point, but there are many other types.

We shall now prove the *monodromy theorem:*

In a simply connected domain D, the values of the analytic function $f(z)$ obtained by continuation from a single element in D necessarily constitute a single-valued branch of $f(z)$ in D.

Assume the theorem to be false. Then there are two points z_1 and z_2 and two contours starting at z_1 and ending at z_2 lying in D such that the continuation of the same functional element at z_1 along these contours leads to two different elements at z_2. Or, using the closed contour $-C_1 + C_2$, continuation of an element at z_2 around this leads back to a different element. By a construction like that of Section 49, we may find first a closed polygon in D having the same property as $-C_1 + C_2$, then a simple closed polygon, and then a triangle. And by continued subdivision of triangles into four smaller triangles as in Section 48, we may find a succession of triangular contours T_n closing down on a point where each T_n has this same property. Call the point which is in all the triangles T_n, z_0. At z_0 the Taylor's series for $f(z)$ will converge in some circle C_0. For N sufficiently large, the triangle T_N will lie entirely in C_0. Then continuation around T_N will be effected by a single functional element, and so cannot lead back to an element different from the one we started with. Since the assumption that the theorem was false led to a contradiction, we have proved the monodromy theorem to be true.

Suppose that some element of an analytic function has an infinite radius of convergence. Then the function is an entire function with no singularities in the open plane. And the element at any point has an infinite radius of convergence and may be used as a global definition of the function.

If a function is not an entire function, each of its elements has a finite radius of convergence. The following theorem gives some information about its singularities.

If a function element has a finite radius of convergence, there must be at least one singular point on C, the circle of convergence, $|z - a| = R$.

Suppose that this were not the case. Then by analytic continuation we could obtain a function element for each point z on C, with a circle of convergence C'_z. Since these circles cover C, by the Heine-Borel theorem a finite subset would also cover C. This would lead to a simply connected domain D containing all points with $|z - a| \leq R'$, for some $R' > R$. By the monodromy theorem, the continuation of $f(z)$ into D would be single-valued as well as analytic. This contradicts the description of R in Corollary 3 of Section 53. Thus the assumption of no singularity on C is untenable, as we set out to prove.

Let z_1 be any point inside the circle C, and let C_1 be the circle of convergence of the element at z_1 in powers of $(z - z_1)$. Let K_1 denote the circle with center z_1 which touches C internally at the point where $z = b$. Hence K_1 has its radius equal to $R - |z_1 - a|$. The element at z_1 is necessarily convergent within K_1 and has the same sum as the element at a there. Hence the radius of C_1 cannot be less than that of K_1. If the radius of C_1 exceeds that of K_1, the second element continues the function along the radius from a to b beyond b, and the function is analytic at b. But if the radius of C_1 equals that of K_1, the point b is a singularity of $f(z)$, for there must be one singularity on the circumference of C_1, and b is the only point on C_1 not lying inside C, where the function is analytic.

It is possible for every point of C to be a singularity of $f(z)$. In this case the circle C is a *natural boundary* of the function. See Problem 11 of Exercise 29.

Sec. 60   TAYLOR'S EXPANSION   175

Exercise 29

1. Assuming that the addition theorem for the sine,
$$\sin(z+w) = \sin z \cos w + \cos z \sin w,$$
is known to be true for real values of z and w, prove that it must also hold for all complex values of z and w. *Hint:* Starting with z and w real, first prove the theorem for z complex, w real by analytic continuation on z. Then prove it for z complex, w complex by analytic continuation on w.

2. A monogenic function has as one element
$$f(z) = z - \frac{z^2}{2} + \frac{z^3}{3} - \ldots + (-1)^{n+1}\frac{z^n}{n} + \ldots, \quad |z| < 1.$$
Prove that for all analytic continuations of this element and for all values of z, its derivative is the single-valued function $f'(z) = 1/(1+z)$. *Hint:* From the series, deduce the result for $|z| < 1$, and then use the principle of permanence of form.

3. Show that for b real, the monogenic function of Problem 2 has
$$F(z) = \frac{1}{2}\log(1+b^2) + i\tan^{-1}b + \frac{z-ib}{1+ib}$$
$$- \frac{1}{2}\left(\frac{z-ib}{1+ib}\right)^2 + \frac{1}{3}\left(\frac{z-ib}{1+ib}\right)^3 - \ldots$$
as one of its elements. *Hint:* From the series for $F(z)$, deduce that for
$$\left|\frac{z-ib}{1+ib}\right| < 1, \quad F'(z) = \frac{1}{1+z}.$$
Also, from Problem 2,
$$f(ib) = \int_0^b \frac{i\,dy}{1+iy} = \int_0^b \frac{y\,dy}{1+y^2} + i\int_0^b \frac{dy}{1+y^2}$$
for the given element if $|b| < 1$, and for the continuation along the real axis if $|b| > 1$.

4. Show that the element $f(z)$ of Problem 2 has no domain of convergence in common with that of the element
$$G(z) = i\pi - (z+2) - \tfrac{1}{2}(z+2)^2 - \tfrac{1}{3}(z+2)^3 + \ldots,$$
but that $G(z)$ is an analytic continuation of $f(z)$. *Hint:* From the series for $G(z)$ deduce that for $|z+2| < 1$, $G'(z) = 1/(1+z)$. Also,

from Problem 2, for the continuation along the upper semicircle of $z = -1 + e^{i\theta}$, we have

$$f(-2) = \int_0^\pi i\, d\theta = i\pi.$$

5. Check Problems 3 and 4 by deducing that each of the elements $f(z)$, $F(z)$, $G(z)$ is equal to a branch of $\log(1 + z)$.

That the same analytic *expression* can represent two different analytic *functions* has been illustrated in Problems 7 to 9 of Exercise 26, where the expressions involved integrals. Verify the following additional illustrations involving limits or series.

6. $\lim\limits_{n \to \infty} \dfrac{e^z + z^{n+1}}{1 + z^n} = e^z$ when $|z| < 1$, but $= z$ when $|z| > 1$.

7. $\dfrac{z}{1 - z^2} + \dfrac{z^2}{1 - z^4} + \dfrac{z^4}{1 - z^8} + \cdots + \dfrac{z^{2^{n-1}}}{1 - z^{2^n}} + \cdots = \dfrac{z}{1 - z}$

when $|z| < 1$, but $= \dfrac{1}{1 - z}$ when $|z| > 1$. *Hint:* Show that the nth term equals

$$\frac{1}{1 - z^{2^{n-1}}} - \frac{1}{1 - z^{2^n}},$$

and that the sum to n terms equals

$$\frac{1}{1 - z} - \frac{1}{1 - z^{2^n}}.$$

8. Show that in the monodromy theorem the hypothesis that D is simply connected is necessary, by verifying that $f(z) = \log z$ has analytic elements in the domain $0 < |z| < 2$, but that the continuation from the element at $z = 1$,

$$\text{Log } z = (z - 1) - \frac{(z - 1)^2}{2} + \frac{(z - 1)^3}{3} - \cdots$$

around the circle $z = e^{i\theta}$ from $\theta = 0$ to $\theta = 2\pi$ must lead to the element of $\log z$ at $z = 1$ which equals

$$2\pi i + (z - 1) - \frac{(z - 1)^2}{2} + \frac{(z - 1)^3}{3} - \cdots.$$

Hint: For all elements, $f'(z) = 1/z$, so that we return to

$$\text{Log } z + \int_0^{2\pi} i\, d\theta = \text{Log } z + 2\pi i.$$

9. Let $f(z) = z - \frac{z^2}{2^2} + \frac{z^3}{3^2} - \ldots + (-1)^{n+1}\frac{z^n}{n^2} + \ldots$ for $|z| < 1$.

Show that the series converges for *all* $|z| \leq 1$, but that the only singular point on the circle of convergence is at $z = -1$. *Hint:* As in Problem 2, derive

$$f''(z) = \frac{1}{1+z}, \quad f'(z) = \text{Log}(1+z),$$

and

$$f(z) = (1+z)\text{Log}(1+z) - z, \text{ for } |z| < 1.$$

10. Verify that for the function $1/z$, the element at $z \neq 0$ is

$$\frac{1}{a} + \frac{(z-a)}{a^2} + \frac{(z-a)^2}{a^3} + \ldots + \frac{(z-a)^n}{a^{n+1}} + \ldots$$

Also show that to define the function for all values of $z \neq 0$, we need an infinite number of elements, but that for any finite domain three elements suffice. *Hint:* The domain $|z| < r$ is completely covered by the circles of convergence for $z_1 = r$, $z_2 = re^{2\pi i/3}$, $z_3 = re^{-2\pi i/3}$.

11. Verify that every point of the circle of convergence of the element

$$f(z) = z + z^2 + z^4 + \ldots + z^{2^n} + \ldots \text{ for } |z| < 1$$

is a singular point. A curve of this type across which continuation is impossible is called a *natural boundary* of the function. *Hint:* For z equal to a real number r, the sum of the series can be made to exceed any given number N by taking r sufficiently near to unity. Hence in every neighborhood of $z = 1$, for any given N, there are points where $|f(z)| > N$. Such values cannot belong to an element regular at $z = 1$. From the series, $f(z) = z + f(z^2)$, so that there is similar behavior at the two values with $z^2 = 1$, the four values with $z^4 = 1$, the 2^n values with $z^{2^n} = 1$, and the limit points of these fill up the circle $z = 1$.

12. Analytic continuation may sometime be practically carried out by the following *principle of reflection*. Let the function $f(z)$ be single-valued and analytic in a closed region R_1 in the upper half plane. Let R_1 have a segment AB of the real axis as part of its boundary. Also let the values of $f(z)$ be real and continuous on AB. Let R_2 be the region in the lower half plane which is the reflection of R_1 in the real axis, so that if z is in R_1, the conjugate value $\bar{z}$ is in R_2. Then $\overline{f(z)}$ defines the continuation of $f(z)$ from R_1 to R_2. Prove this. *Hint:* Use

the Cauchy-Riemann equations to show that $\overline{f(\bar{z})}$ is analytic in R_2. Then use the form of the Cauchy-Goursat theorem of Corollary 2 of Section 51, combined with a cross cut along part of AB if necessary, to show that $F(z) = f(z)$ in R_1 and $\overline{f(\bar{z})}$ in R_2 satisfies the conditions of Morera's theorem.

9

LAURENT'S EXPANSION

Consider a function $f(z)$ which is single-valued and analytic in a concentric annular ring $r_1 < |z - a| < r_2$. Nothing need be known about the behavior of the function at points outside the ring. We shall show that inside the ring the function $f(z)$ may be represented by an expansion involving positive and negative powers of $(z - a)$. This Laurent expansion is unique for a given function and annular ring, and we obtain integral expressions for its coefficients. Such expansions are particularly important when a is an isolated singularity of $f(z)$, and we take as the ring some deleted neighborhood of a, $0 < |z - a| < r_2$. In this case the Laurent series may be used to study the nature of the singular point, and the behavior of the function near a. We show that isolated singularities are either removable by a suitable definition of $f(a)$, poles near which $f(z)$ becomes infinite, or essential singularities near which every complex value is approximately assumed.

We develop a number of properties of rational functions, and in particular show that they are characterized by being single-valued and having no other singularities than poles in the closed plane.

The integral $\int_C f(z)\, dz$ where C is a sufficiently small positive circuit about a is zero when $f(z)$ is analytic at a. But when a is an isolated singularity, or is the point at infinity, this integral is in general different from zero. We use it to define the residue of $f(z)$ at a as the integral divided by $2\pi i$. This leads to the residue theorem on the evaluation of certain integrals in terms of residues, and also to the amplitude principle which provides an expression for the number of zeros minus the number of poles of a function in a given domain, account being taken of multiplicities. We apply this to Rouché's theorem on the zeros of functions as well as to a result on the one-to-one nature of certain conformal maps. And because it is related to this we describe the Schwarz-Christoffel transformation which enables us to map the interior of a polygon conformally on a half plane.

61. Laurent's series

Let D be the annular domain between two concentric circles, $r_1 < |z - a| < r_2$. Let C_1 denote the inner circle $|z - a| = r_1$, and let C_2 denote the outer circle $|z - a| = r_2$, each traversed positively. Thus $C_2 - C_1$ is the complete properly oriented boundary of the multiply connected domain D. Consider a function $f(z)$ which is single-valued and analytic for $r_1 \leq |z - a| \leq r_2$ or in D and on C_1 and C_2. Then by the theorem on Cauchy's integral formula of Section 51, for z in D, we have

$$f(z) = \frac{1}{2\pi i} \int_{C_2} \frac{f(t)\, dt}{t - z} - \frac{1}{2\pi i} \int_{C_1} \frac{f(t)\, dt}{t - z}. \tag{1}$$

For a fixed z inside D, and a variable t on C_2, we have

$$|z - a| < r_2 \quad \text{and} \quad \left|\frac{z - a}{t - a}\right| = \frac{|z - a|}{r_2} < 1. \tag{2}$$

And by the argument which led from Eq. (5) to Eqs. (10) and (11) of Section 53, we may conclude that

$$f_2(z) = \frac{1}{2\pi i} \int_{C_2} \frac{f(t)\, dt}{t - z}$$
$$= A_0 + A_1(z - a) + A_2(z - a)^2 + \ldots + A_n(z - a)^n + \ldots, \tag{3}$$

where

$$A_n = \frac{1}{2\pi i} \int_{C_2} \frac{f(t)\, dt}{(t - a)^{n+1}}. \tag{4}$$

LAURENT'S EXPANSION

For a fixed z inside D, and a variable t on C_1, we have

$$r_1 < |z - a| \quad \text{and} \quad \left|\frac{t-a}{z-a}\right| = \frac{r_1}{|z-a|} < 1. \tag{5}$$

From Eq. (6) of Section 53, with t and z interchanged, we may deduce that

$$\frac{1}{z-t} = \frac{1}{z-a} + \frac{t-a}{(z-a)^2} + \frac{(t-a)^2}{(z-a)^3} +$$
$$\cdots + \frac{(t-a)^{n-1}}{(z-a)^n} + \frac{(t-a)^n}{(z-a)^n(z-t)}. \tag{6}$$

With t_1 the point on C_1 nearest to z, we have for the remainder,

$$\left|\frac{(t-a)^n}{(z-a)^n(z-t)}\right| \leq \left[\frac{r_1}{|z-a|}\right]^n \frac{1}{|z-t_1|}. \tag{7}$$

The right member is independent of t, and tends to zero when n becomes infinite. It follows that the series

$$\frac{1}{z-a} + \frac{t-a}{(z-a)^2} + \frac{(t-a)^2}{(z-a)^3} + \cdots + \frac{(t-a)^{n-1}}{(z-a)^n} + \cdots \tag{8}$$

converges to $1/(z-t) = -1/(t-z)$ uniformly with respect to t.

Since $f(t)$ is bounded on C_1, termwise multiplication of the series (8) by $f(t)$ will not disturb the uniform convergence. By Section 52 we may integrate the new series termwise, to obtain

$$f_1(z) = -\frac{1}{2\pi i} \int_{C_1} \frac{f(t)\,dt}{t-z}$$
$$= \frac{A_{-1}}{z-a} + \frac{A_{-2}}{(z-a)^2} + \frac{A_{-3}}{(z-a)^3} + \cdots + \frac{A_{-n}}{(z-a)^n} + \cdots, \tag{9}$$

where

$$A_{-n} = \frac{1}{2\pi i}\int_{C_1} (t-a)^{n-1} f(t)\,dt. \tag{10}$$

The integrands in Eqs. (4) and (10) have the same form, since

$$F_k(t) = \frac{f(t)}{(t-a)^{k+1}} = (t-a)^{-k-1} f(t) \tag{11}$$

equals the integrand in Eq. (4) for $k = n$, zero or a positive integer, and equals the integrand in Eq. (10) for $k = -n$, a negative integer. Since the factor $(t-a)^{-k-1}$ is analytic for $t \neq a$, our hypothesis on $f(t)$

makes $F_k(t)$ analytic for $r_1 \leq |t - a| \leq r_2$. Let C_3 denote any closed curve in D making a single positive circuit about a. Then $C_2 - C_3$ and $C_3 - C_1$ are each the complete properly oriented boundary of a multiply connected domain, such that $F_k(t)$ is analytic in the domain and on its boundary. Hence by the Cauchy-Goursat theorem of Corollary 2, Section 51, we must have

$$\int_{C_2 - C_3} F_k(t)\, dt = 0, \quad \int_{C_3 - C_1} F_k(t)\, dt = 0. \tag{12}$$

This implies that

$$\int_{C_2} F_k(t)\, dt = \int_{C_3} F_k(t)\, dt = \int_{C_1} F_k(t)\, dt. \tag{13}$$

Thus in Eqs. (4) and (10) we may use the circle C_1, the circle C_2, or any closed curve C_3. This proves *Laurent's theorem*:

Let $f(z)$ be single-valued and analytic for $r_1 \leq |z - a| \leq r_2$. Then the Laurent expansion

$$f(z) = \sum_{k=-\infty}^{\infty} A_k (z - a)^k \tag{14}$$

is valid for $r_1 < |z - a| < r_2$. And the coefficients A_k are given by

$$A_k = \frac{1}{2\pi i} \int_C \frac{f(t)\, dt}{(t - a)^{k+1}} \tag{15}$$

with C any single positive circuit about a, in particular either boundary circle, made up entirely of points with $r_1 \leq |z - a| \leq r_2$.

Corollary 1. The conclusions of the theorem, Eqs. (14) and (15), follow from the mere assumption that $f(z)$ is single-valued and analytic in the open annulus $r_1 < |z - a| < r_2$, if in Eq. (15) C is restricted to be made up entirely of points with $r_1 < |z - a| < r_2$.

To prove this, let C be any such curve and z_0 be any point with $r_1 < |z_0 - a| < r_2$. Then we may find an R_1 and an R_2 such that $r_1 < R_1 < R_2 < r_2$, all the points of C are such that $R_1 < |z - a| < R_2$, and also $R_1 < |z_0 - a| < R_2$. We then apply the theorem to the closed annulus $R_1 \leq |z - a| \leq R_2$ to show that the series converges at z_0, which was any point of the open annulus $r_1 < |z - a| < r_2$.

Corollary 2. If $f(z)$ is single-valued and analytic for $r_1 < |z - a| < r_2$, then $f(z) = f_1(z) + f_2(z)$, where $f_1(z)$ is analytic for $|z - a| > r_1$, including infinity, while $f_2(z)$ is analytic for $|z - a| < r_2$.

To show this, we consider the expansion of Eq. (14), which is valid for $r_1 < |z - a| < r_2$, by Corollary 1. Let

$$f_2(z) = \sum_{n=0}^{\infty} A_n (z-a)^n, \quad f_1(z) = \sum_{n=1}^{\infty} A_{-n}(z-a)^{-n}. \qquad (16)$$

First consider any z with $|z - a| < r_2$. If $z = z''$, with $r_1 < |z'' - a|$, the series for $f_2(z)$ converges since z'' is in the open annulus. But if $z = z'$, with $|z' - a| \leq r_1$, then $|z' - a| < |z'' - a|$ and the power series for $f_2(z)$ converges by Theorem 1 of Section 19. Hence the power series for $f_2(z)$ converges for $|z - a| < r_2$ so that $f_2(z)$ is analytic for these values as stated in the corollary.

Next consider any z with $|z - a| > r_1$. If $z = z''$ with $|z'' - a| < r_2$, the series for $f_1(z)$ converges since z'' is in the open annulus. That is, the power series $\Sigma A_{-n} Z^n$ converges for $Z'' = 1/(z'' - a)$. Hence by Theorem 1 of Section 19, it converges for any Z' with $|Z'| < |Z''|$, or in particular for $Z' = 1/(z' - a)$ if $|z' - a| \geq r_1 > |z'' - a|$. Hence the power series $\Sigma A_{-n} Z^n$ is an analytic function of Z for $|Z| < 1/r_1$. With $Z = 1/(z - a)$, this makes $f_1(z)$ an analytic function of z for $|z - a| > r_1$ as stated in the corollary.

Corollary 3. Let the function $f(z)$ be single-valued and analytic at each point in a simply connected domain D with the possible exception of the point a in D. Choose an r_2 such that all the points with $|z - a| < r_2$ are in D. Then the Laurent expansion of Eq. (14) is valid for $0 < |z - a| < r_2$, and the coefficients A_k are given by Eq. (15) with C any single positive circuit about a made up entirely of points in D.

Corollary 4. The Laurent expansion of $f(z)$ for a particular annular domain is unique in the following sense. Let

$$\sum_{k=-\infty}^{\infty} B_k(z-a)^k = \sum_{k=-\infty}^{\infty} A_k(z-a)^k = f(z) \qquad (17)$$

for z in D, or $r_1 < |z - a| < r_2$. Then $B_k = A_k$ for all values of k, and the common sum function $f(z)$ is analytic for all z in D.

We prove this as follows. From the mere convergence of the two series in D, by an argument based on Theorem 1 of Section 19 like that used for Corollary 2, we may deduce that each of the two power series

$$\sum_{n=0}^{\infty} A_n(z-a)^n = f_2(z) \quad \text{and} \quad \sum_{n=0}^{\infty} B_n(z-a)^n = F_2(z) \quad (18)$$

converges for $|z-a| < r_2$, and each of the two power series

$$\sum_{n=1}^{\infty} A_n Z^n = f_1(z) \quad \text{and} \quad \sum_{n=1}^{\infty} B_n Z^n = F_1(z) \quad \text{with } Z = \frac{1}{z-a} \quad (19)$$

converges for $|Z| < 1/r_1$.

Next let $r_3 = \frac{1}{2}(r_1 + r_2)$, so that $r_1 < r_3 < r_2$. Then from Theorem 3 of Section 20 it follows that the two power series of Eq. (18) converge uniformly for $|z - a| \leq r_3$, while the two power series of Eq. (19) converge uniformly for $|Z| \leq 1/r_3$, or for $|z - a| \geq r_3$. In particular, all four series converge uniformly on the circle C_3: $|z - a| = r_3$. But on C_3, $|(z-a)^{-k-1}| = r_3^{-k-1}$ is constant, and hence bounded. Thus termwise multiplication of the series in Eq. (17) by $(z-a)^{-k-1}$ will not disturb the uniform convergence. And by Section 52, after such multiplication we may integrate termwise over C_3. By Problem 11 of Exercise 24, all the terms give zero except $A_k(z-a)^{-1}$ and $B_k(z-a)^{-1}$. Thus we find

$$B_k = A_k = \frac{1}{2\pi i} \int_{C_3} \frac{f(t)\, dt}{(t-a)^{k+1}}. \quad (20)$$

This proves that $B_k = A_k$ for all values of k. The integral form agrees with Eq. (15), if we take C_3 as the circuit C in that equation.

Since the power series of Eq. (18) converge for all z such that $|z - a| < r_2$, they are analytic for such values. Similarly the power series of Eq. (19) converge for all Z with $|Z| < 1/r_1$, and so are analytic functions of Z for these values, or analytic functions of z for $|z - a| > r_1$. Thus $f_1(z)$ and $f_2(z)$ are each analytic for $r_1 < |z - a| < r_2$, or in D. This makes $f(z) = f_1(z) + f_2(z)$ analytic in D, the last conclusion of Corollary 4.

Exercise 30

1. Prove that, with the hypothesis of the theorem that $f(z)$ is analytic for $r_1 \leq |z - a| \leq r_2$, the Laurent expansion of Eq. (14) is

valid for all these values, that is, not only in the open annulus, but for the boundary as well. *Hint:* By an argument like that of Section 51, use the analyticity on the boundary and the Heine-Borel theorem to obtain a slightly larger open annulus, $r_1' < |z - a| < r_2'$ with $r_1' < r_1, r_2' > r_2$, in which $f(z)$ is analytic.

2. Use the binomial theorem, Problem 9 of Exercise 27, to deduce the finite Laurent expansion,

$$\left(z + \frac{1}{z}\right)^{2N} = \frac{(1 + z^2)^{2N}}{z^{2N}}$$

$$= \sum_{m=-N}^{N} \frac{2N(2N-1)(2N-2)\ldots(N-m+1)}{(m+N)!} z^{2m}.$$

3. From Eq. (15) with $f(z) = (z + 1/z)^{2N}$, $a = 0$, and C the unit circle with $z = e^{i\theta}$, deduce that

$$A_0 = \frac{1}{2\pi i} \int_C \left(z + \frac{1}{z}\right)^{2N} \frac{dz}{z} = \frac{1}{2\pi} \int_0^{2\pi} 2^{2N} \cos^{2N} \theta \, d\theta.$$

Use this and the value of A_0 from Problem 2 to prove that

$$\int_0^{\pi/2} \sin^{2N} \theta \, d\theta = \int_0^{\pi/2} \cos^{2N} \theta \, d\theta = \frac{\pi}{2} \frac{1}{2^{2N}} \frac{(2N)!}{N!N!}$$

$$= \frac{\pi}{2} \frac{1 \cdot 3 \cdot 5 \ldots (2N-1)}{2 \cdot 4 \cdot 6 \ldots (2N)}.$$

Using known Taylor's series, deduce that each of the following Laurent expansions is valid for $z \neq 0$.

4. $\left(\dfrac{1}{z^2} + \dfrac{2}{z}\right) e^z = \dfrac{1}{z^2} + \dfrac{3}{z} + \dfrac{5}{2} + \ldots + \dfrac{2n+5}{(n+2)!} z^n + \ldots$

5. $\cos \dfrac{1}{z} + \cos z - 1 = \sum_{m=-\infty}^{\infty} (-1)^m \dfrac{z^{2m}}{m!}.$

6. $\left(\dfrac{6}{z^3} + \dfrac{1}{z}\right) \sin z = \dfrac{6}{z^2} - \dfrac{14}{5!} z^2 +$

$$\ldots + (-1)^m \frac{4m^2 + 10m}{(2m+3)!} z^{2m} + \ldots.$$

From Problem 7 of Exercise 27, by interchanging Z and b, deduce that

7. $\dfrac{1}{z-b} = -\dfrac{1}{b} - \dfrac{z}{b^2} - \dfrac{z^2}{b^3} - \ldots - \dfrac{z^n}{b^{n+1}} - \ldots$ for $|z| < |b|$.

but
$$\frac{1}{z-b} = \frac{1}{z} + \frac{b}{z^2} + \frac{b^2}{z^3} + \ldots + \frac{b^{n-1}}{z^n} + \ldots \text{ for } |z| > |b|.$$

8. $\dfrac{1}{(z-b)^m} = \left(-\dfrac{1}{b}\right)^m \left[1 + \dfrac{mz}{b} + \right.$
$$\left. \ldots + \frac{(n+1)\ldots(n+m-1)}{(m-1)!} \frac{z^n}{b^n} + \ldots \right],$$

where $m > 1$ for $|z| < |b|$, but for $|z| > |b|$ we have
$$\frac{1}{(z-b)^m} = \frac{1}{z^m} + \frac{mb}{z^{m+1}} +$$
$$\ldots + \frac{(n+1)\ldots(n+m-1)}{(m-1)!} \frac{b^n}{z^{m+n}} + \ldots.$$

Use Problem 7 to deduce that

9. $\dfrac{1}{1+z^2} = 1 - z^2 + z^4 - \ldots + (-1)^m z^{2m} + \ldots$ for $|z| < 1$,

but
$$\frac{1}{1+z^2} = \frac{1}{z^2} - \frac{1}{z^4} + \frac{1}{z^6} - \ldots + (-1)^{m-1} \frac{1}{z^{2m}} + \ldots \text{ for } |z| > 1.$$

10. The Laurent expansion of
$$\frac{1}{z^2 - 3z + 2} = \frac{-1}{z-1} + \frac{1}{z-2}$$

is
$$\frac{1}{2} + \frac{3}{4}z + \frac{7}{8}z^2 + \ldots + \left(1 - \frac{1}{2^{n+1}}\right)z^n + \ldots \text{ for } |z| < 1,$$
$$\sum_{n=1}^{\infty} \frac{-1}{z^n} + \sum_{n=0}^{\infty} \frac{-z^n}{2^{n+1}} \text{ for } 1 < |z| < 2, \quad \sum_{n=2}^{\infty} \frac{2^{n-1} - 1}{z^n} \text{ for } |z| > 2.$$

62. Singular points

Consider a function $f(z)$ which is analytic and single-valued in some deleted neighborhood of a, that is, at all points of D with the possible exception of a, where D is some simply connected domain having a as an interior point. Then by Corollary 3 of Section 61, there is a Laurent expansion of the type of Eq. (14).

Suppose first that this expansion contains no negative powers.

Then its sum is a function analytic at a. If $f(a)$ was given as equal to A_0, the function $f(z)$ was analytic at a to begin with. If $f(a)$ was undefined, or defined as a value different from A_0, then $f(z)$ was not analytic at a, but if we redefine our function at a as A_0 it becomes analytic. In this case the original function is said to have a *removable singularity* at a. Since defining $f(a)$ as A_0 makes the function continuous at a, it follows that if $f(z)$ is analytic in some deleted neighborhood of a, and continuous at a, then $f(z)$ is analytic at a.

Suppose next that the expansion of Eq. (14) contains some, but only a finite number of negative powers. Then it can be written

$$f(z) = \sum_{k=-m}^{\infty} A_k(z-a)^k, \qquad A_{-m} \neq 0. \tag{21}$$

The negative terms in any Laurent expansion which is valid for $0 < |z - a| < r_2$ form a series which is called the *principal part* of $f(z)$ at a. Here the principal part consists of the $(m+1)$ terms

$$\frac{A_{-1}}{z-a} + \frac{A_{-2}}{(z-a)^2} + \cdots + \frac{A_{-m}}{(z-a)^m}. \tag{22}$$

And the function $f(z)$ is said to have a *pole* at the point a, of order m. Thus m, the *order* of the pole, is the highest power of $(z-a)^{-1}$ in the expansion. As a consequence of Eq. (21), the product $f(z)(z-a)^m$ is equal to a power series and so has a removable singularity at a. And the order m is the smallest positive integer for which this is true. We note that

$$\lim_{z \to a} (z-a)^m f(z) = A_{-m} \neq 0. \tag{23}$$

It follows that when $f(z)$ has a pole at a,

$$\lim_{z \to a} |f(z)| = \infty \tag{24}$$

The coefficient of $(z-a)^{-1}$ in Eq. (22), or A_{-1} is called the *residue* of $f(z)$ at the pole $z = a$. For a pole of order one, $A_{-1} = \lim_{z \to a}(z-a)f(z)$, by Eq. (23) with $m = 1$, and $A_{-1} \neq 0$. For $m > 1$, the evaluation of A_{-1} is less simple, and in exceptional cases we may have $A_{-1} = 0$.

If we accept the limiting value from Eq. (23) as the definition of the product at a, then $(z-a)^m f(z)$ is analytic at a and is different from zero there. Hence its reciprocal

$$F(z) = \frac{1}{f(z)(z-a)^m} \tag{25}$$

is analytic at a and is different from zero there.

Let the function $g(z)$ be analytic at a. And in its Taylor's expansion, Eq. (10) of Section 53, let $A_0 = 0$, and A_m be the first nonzero coefficient. Then that expansion may be written

$$g(z) = \sum_{n=m}^{\infty} A_n(z-a)^n, \qquad A_m \neq 0, \, m > 0. \tag{26}$$

Such a function $g(z)$ is said to have a zero of *order m* at $z = a$.

A comparison of Eqs. (25) and (26) shows that, if $f(z)$ has a pole of order m at $z = a$, the function $1/f(z)$ has a zero of order m at $z = a$. And conversely if a function has a zero of order m at $z = a$, its reciprocal has a pole of order m at $z = a$.

If a function is *not* analytic at a, but is single-valued and analytic in some deleted neighborhood of a, then a is called an *isolated singular point*. An isolated singularity which is not removable and not a pole is an *essential singularity*. Consequently at an essential singularity, the principal part of $f(z)$ consists of an infinite series of negative terms. In this infinite series, A_{-1}, the coefficient of $(z-a)^{-1}$, is called the *residue* of $f(z)$ at the essential singularity.

If the Laurent expansion which is valid for $0 < |z - a| < r_2$ is known, we may easily classify the type of singularity at a by examining the principal part. As a first step toward finding the type of singularity from the values of the function, we prove *Riemann's theorem*:

Let $f(z)$ be single-valued and analytic in some deleted neighborhood of a. Then if $f(z)$ is uniformly bounded in this deleted neighborhood, $f(z)$ is either analytic at a or has a removable singularity there.

By Corollary 3 of Section 61 there is a Laurent expansion of the type of Eq. (14) valid for $0 < |z - a| < r_2$. And the coefficients may be found from Eq. (15) with C any circle $|z - a| = r$ with $r < r_2$. Then if $|f(z)| \leq M$ in the deleted neighborhood, $|f(t)| \leq M$ for $|t - a| < r_2$ and hence for t on any circle C. Hence from Eq. (15), and Eq. (48) of Section 46, we may deduce that

Sec. 62 LAURENT'S EXPANSION

$$|A_{-n}| = \left|\frac{1}{2\pi i}\int_C (t-a)^{n-1}f(t)\,dt\right| \leq Mr^n. \tag{27}$$

For any fixed $n > 0$, $Mr^n \to 0$ when $r \to 0$. It follows that the constant coefficient $A_{-n} = 0$ for any $n > 0$. Thus the expansion of Eq. (14) contains no negative powers, and $f(z)$ is either analytic at a or has a removable singularity there, as the theorem asserts.

If the values of $f(z)$ do not admit an upper bound M in any deleted neighborhood of the isolated singularity a, there must be a sequence of values z_n with

$$\lim_{n\to\infty} z_n = a, \qquad \lim_{n\to\infty} |f(z_n)| = \infty. \tag{28}$$

At a pole, this is true for every sequence of points z_n with $z_n \to a$, by Eq. (24). Conversely, if $|f(z)| \to \infty$ for $z \to a$, the function has a pole at a. For let $G(z) = 1/f(z)$. Then

$$\lim_{z\to a} |G(z)| = \lim_{z\to a}\frac{1}{|f(z)|} = 0. \tag{29}$$

This shows that $G(z)$ is uniformly bounded in some deleted neighborhood of a. Hence by Riemann's theorem it has at most a removable singularity at a and becomes analytic if we accept the limiting value 0 as its value at a. Since $G(z)$ is not zero for $z \neq a$, then $G(z)$ is not identically zero. Hence $G(z)$ has a zero order m at a, and its reciprocal $f(z)$ has a pole of order m at a.

Consider the behavior of the values of $f(z)$ when $z \to a$, an isolated singular point. Approach to the same finite value for all sequences characterizes a removable singularity, while becoming infinite for all sequences characterizes a pole. The extent to which distinct behavior for different sequences must occur when a is an essential singularity is indicated by *Weierstrass's theorem:*

If a is an isolated essential singularity, for a suitably chosen sequence of values $z_n \to a$, $f(z_n)$ can be made to become infinite, or to approach any given finite value whatever.

There must be a sequence satisfying Eq. (28), or $f(z)$ would admit an upper bound in some deleted neighborhood of a, and $f(z)$ would have only a removable singularity at a.

Let b be any given finite complex number. Form the function

$1/[f(z) - b]$, and consider its behavior near a. If it approached the same finite value or became infinite for all sequences $z_n \to a$, the behavior of $f(z)$ would be correspondingly restricted. Thus this function $1/[f(z) - b]$ must also have an essential singularity at a. And by the preceding paragraph, there is a sequence z_n such that

$$\lim_{n \to \infty} z_n = a \quad \text{and} \quad \lim_{n \to \infty} \left| \frac{1}{f(z_n) - b} \right| = \infty. \tag{30}$$

This implies that $|f(z_n) - b| \to 0$ and $f(z_n) \to b$. This completes the proof of Weierstrass's theorem.

In accordance with the principle stated in Section 37, $f(z)$ has an isolated singularity at I or $z = \infty$ if $f(1/Z)$ has an isolated singularity at $Z = 0$. Thus there is a Laurent expansion

$$f\left(\frac{1}{Z}\right) = \sum_{k=-\infty}^{\infty} B_{-k} Z^k, \quad \text{with} \quad \sum_{n=1}^{\infty} \frac{B_n}{Z^n} \tag{31}$$

as its principal part, valid for $0 < |Z| < r_2$. Hence

$$f(z) = \sum_{k=-\infty}^{\infty} B_k z^k, \quad \text{with} \quad \sum_{n=1}^{\infty} B_n z^n \tag{32}$$

as its principal part, valid for $|z| > 1/r_2$. Here $f(z)$ is analytic in the deleted neighborhood of I, $|z| > 1/r_2$ with the possible exception of I itself. We say that the isolated singularity at infinity is removable, a pole, or an essential singularity according as the number of nonzero terms in the principal part is zero, finite and positive, or infinite.

For reasons which will be explained in Section 67, the *residue* at *infinity* is defined as the *negative* of the coefficient of $1/z$ in the expansion of Eq. (32). This residue, $-B_{-1}$ can be found either from the term $B_{-1}Z$ in Eq. (31), or from the term B_{-1}/z in Eq. (32). Note that here these terms are *not* contained in the principal part.

We shall now prove a third theorem on the relation of the type of singularity of $f(z)$ at a to special behavior of the values of $f(z)$ near a.

Let $f(z)$ be single-valued and analytic for $0 < |z - a| < r$. And for some infinite sequence of values $z_n \to a$, let $f(z_n) = b$. Then either $f(z) = b$ is constant for $0 < |z - a| < r$, or $f(z)$ has an essential singularity at a.

Since the limit for $f(z)$ as $z \to a$ is finite for one sequence, $f(z)$ cannot have a pole at a.

If $f(z)$ had a removable singularity at a, giving $f(z)$ the value b would lead to a function analytic at a, with $f(z_n) = b$. By Problem 13 of Exercise 27 this must have the same Taylor's series in powers of $(z - a)$ as the function $f(z) = b$, with $A_0 = b$ and $A_n = 0$ for $n > 0$. Hence $f(z) = b$ is constant for $0 < |z - a| < r$, the first alternative.

When this is not the case, $f(z)$ has an isolated singularity at a which is not a pole and not removable. Hence it must be an essential singularity, the other possibility.

Corollary 1. Let $f(z)$ be single-valued and analytic, but not constant, in some deleted neighborhood of a. Then if a is a limit point of zeros of $f(z)$, it must be an essential singularity of $f(z)$.

By the definition of a limit point, and the construction indicated in Section 10, we may find a sequence of zeros of $f(z)$, z_n, such that $z_n \neq a$, and as $n \to \infty$, $z_n \to a$. Then $f(z_n) = 0$, so that we may apply the theorem with $b = 0$. Since here $f(z)$ is not constant, it must have an essential singularity at a.

Corollary 2. Let $f(z)$ be single-valued and analytic for $|z| > r$. And for some infinite sequence of values $z_n \to \infty$ let $f(z_n) = b$. Then either $f(z) = b$ is constant for $|z| > r$, or $f(z)$ has an essential singularity at infinity.

We have merely to apply the theorem to $f(1/Z)$, for $0 < |Z| < 1/r$.

Finally let us consider a function $f(z)$ which is single-valued and analytic at each point of some deleted neighborhood of a with the exception of a single infinite sequence of points a_n, where $a_n \neq a$, but as $n \to \infty$, $a_n \to a$. And let each of these points a_n be an isolated singularity, but not a removable singularity. Then any neighborhood of any one of the points a_n contains an infinite sequence of points on which the values of $f(z)$ become infinite. But any deleted neighborhood of a contains points a_n as interior points, and therefore includes such sequences. It follows that $f(z)$ is not analytic in any deleted neighborhood of a. Thus $f(z)$ cannot be analytic at a, or have a pole or other isolated singularity at a. We call the point a a *nonisolated essential singularity.* We say that $f(z)$ has a *nonisolated essential singularity* for $z = \infty$, if $f(1/Z)$ has such a point for $Z = 0$. That the Weierstrass theorem applies to nonisolated essential singularities is proved in Problem 11 of Exercise 31.

63. Rational functions

Throughout this section, we shall use $f(z)$ to denote a function which is *single-valued* and *analytic* for all points of the closed plane with the exception of the explicitly mentioned isolated and nonremovable singular points. We shall see that rational functions may be characterized by suitable restrictions on their singularities. Our first theorem is

If $f(z)$ has no singularities in the closed plane, it is a constant.

Since $f(z)$ is analytic for $z = \infty$, then $f(1/Z)$ is analytic for $Z = 0$. Hence $|f(1/Z)| < M_1$ for $|Z| < r_1$, and $|f(z)| < M_1$ for $|z| > 1/r_1$. Let M_2 be the maximum value of the continuous real function $|f(z)|$ in the closed region $|z| \leq 1/r_1$. Then $|f(z)| < M_1 + M_2$ for all z. The conclusion that $f(z)$ is a constant follows from Liouville's theorem.

As an analogous result, we have the second theorem:

If $f(z)$ has no singularities in the finite plane, and a pole of order m at infinity, then $f(z)$ is a polynomial of degree m.

Here $f(1/Z)$ has a pole of order m at $Z = 0$, so that the expansion of Eq. (31) has a principal part of the form

$$P\left(\frac{1}{Z}\right) = \frac{B_1}{Z} + \frac{B_2}{Z^2} + \ldots + \frac{B_m}{Z^m}, \quad B_m \neq 0. \tag{33}$$

Let $g(1/Z) = f(1/Z) - P(1/Z)$ for $Z \neq 0$, and $g(1/Z) = B_0$ for $Z = 0$. Then

$$g\left(\frac{1}{Z}\right) = \sum_{k=0}^{\infty} B_{-k} Z^k \quad \text{for} \quad 0 \leq |Z| < r_1. \tag{34}$$

Thus $g(1/Z)$ is analytic for $Z = 0$. This shows that $g(z)$ is analytic for $z = \infty$. But since the polynomial $P(z)$ is analytic for all finite values of z, so is $g(z) = f(z) - P(z)$ for z finite. Hence $g(z)$ is analytic throughout the closed plane, and hence is a constant by our first theorem. Thus $g(z) = B_0$ for all z and

$$f(z) = g(z) + P(z) = B_0 + B_1 z + B_2 z^2 +$$
$$\ldots + B_m z^m, \quad B_m \neq 0 \qquad (35)$$

for all finite z. This proves the second theorem.

We now prove the third theorem:

If $f(z)$ has no essential singularities in the closed plane, then $f(z)$ is a rational function.

If there were an infinite number of isolated singularities in the closed plane, either these would have a finite limit point a, or there would be such points in every neighborhood of I. Hence, by the definition at the end of Section 62, there would be a nonisolated essential singularity either at $z = a$ or at $z = \infty$. Thus $f(z)$ can have as singularities at most a finite number of poles. Let the poles at finite points be $a_1, a_2, \ldots, a_S$. The principal part at a_s will have the form

$$P_s(z) = \frac{A_{-1s}}{z - a_s} + \frac{A_{-2s}}{(z - a_s)^2} + \ldots + \frac{A_{-m_s s}}{(z - a_s)^{m_s}}. \qquad (36)$$

At infinity, the principal part will have the form

$$P_0(z) = B_1 z + B_2 z^2 + \ldots + B_{m_0} z^{m_0} \qquad (37)$$

if $z = \infty$ is a pole. We set $P_0(z) = 0$ if $z = \infty$ is not a pole. Now consider the function

$$g(z) = f(z) - \sum_{s=0}^{S} P_s(z), \qquad (38)$$

for $z \neq a_s$, and define $g(z)$ at a_s and at $z = \infty$ as its limiting finite values at these points. Then $g(z)$ is analytic at all points of the closed plane, and so is a constant B_0, by our first theorem. Hence for $z \neq a_s$, we have

$$f(z) = B_0 + \sum_{s=0}^{S} P_s(z), \qquad (39)$$

a rational function. This proves the third theorem.

Exercise 31

1. Let $f(z)$ and $g(z)$ each have a zero of order m at a. Then they have Taylor's expansions

$$f(z) = \sum_{n=m}^{\infty} A_n(z-a)^n \quad \text{with } A_m \neq 0$$

and

$$g(z) = \sum_{n=m}^{\infty} B_n(z-a)^n \quad \text{with } B_m \neq 0,$$

each valid for $|z-a| < r$ for some suitable r. Show that $f(z)/g(z)$ has a removable singularity at $z = a$, and that the limiting value at a is $A_m/B_m = f^{(m)}(a)/g^{(m)}(a)$.

2. Illustrations of Problem 1 are the function $\sin z/z$ with limiting value 1 at $z = 0$, and the function $(z - \sin z)/z^3$ with limiting value $\frac{1}{6}$ at $z = 0$. Verify that the Taylor's expansions of these quotients are, respectively,

$$1 - \frac{z^2}{3!} + \frac{z^4}{5!} - \cdots \quad \text{and} \quad \frac{1}{3!} - \frac{z^2}{5!} + \frac{z^4}{7!} - \cdots,$$

each valid for $|z| < \infty$.

3. Let

$$f(z) = \sum_{n=0}^{\infty} A_n(z-a)^n \quad \text{with } A_0 = f(a) \neq 0,$$

and

$$g(z) = \sum_{n=m}^{\infty} B_n(z-a)^n \quad \text{with } B_m \neq 0 \text{ for } |z-a| < r.$$

Show that $f(z)/g(z)$ has a pole of order m at a. Also show that $G(z) = (z-a)^m f(z)/g(z)$ has a removable singularity at a. And by considering its Taylor's expansion, deduce that the residue of $f(z)/g(z)$ at a is $\dfrac{1}{(m-1)!} G^{(m-1)}(a)$ for $m > 1$, and is $G(a)$ for $m = 1$.

In Problem 3, show that the residue is

4. $\dfrac{A_0}{B_1} = \dfrac{f(a)}{g'(a)}$ for $m = 1$.

5. $\dfrac{A_1 B_2 - A_0 B_3}{B_2^2} = \dfrac{6f'(a)g''(a) - 2f(a)g'''(a)}{3[g''(a)]^2}$ for $m = 2$.

6. Show that the function e^z has an isolated essential singularity at infinity. Illustrate Weierstrass's theorem by noting that, for any $b_1 = r_1 e^{i\theta_1} \neq 0$, $e^z = b_1$ at all points of the sequence $z_n = \operatorname{Log} r_1 + (\theta_1 + 2n\pi)i$, and that $z_n \to \infty$ as $n \to \infty$. Use $z_n = \operatorname{Log} n$ to give a sequence with $\lim e^{z_n} = \infty$.

7. Show that the function $e^{1/z}$ has an isolated essential singularity at $z = 0$ with a Laurent expansion valid for $0 < |z|$,

$$e^{1/z} = 1 + \frac{1}{z} + \frac{1}{2!z^2} + \ldots + \frac{1}{n!z^n} + \ldots.$$

Exhibit sequences $z_n \to 0$ such that $\lim e^{1/z_n} = \infty$, 0, or any finite value $b_i \neq 0$. *Hint:* Put $z = 1/Z$ and use Problem 6.

8. Show that e^z has a residue zero at the essential singularity at infinity discussed in Problem 6. Also show that $f(z) = 3/z$ is analytic at infinity, but has a residue there equal to -3.

9. Verify that infinity is an isolated essential singularity which is the limit of zeros, $z_n = n\pi$, for $\sin z$. Also verify that $\sec z = 1/(\cos z)$ has a nonisolated essential singularity at infinity which is the limit of poles at $z'_n = n\pi/2$, while $\tan z$ has one which is the limit of both zeros at z_n and poles at z'_n.

10. By putting $z = 1/Z$ in Problem 9, deduce that at the origin $\sin (1/z)$ has an isolated essential singularity which is the limit of zeros, $\sec (1/z)$ has a nonisolated essential singularity which is the limit of poles, while $\tan (1/z)$ has a nonisolated essential singularity which is the limit of both zeros and poles.

11. Prove that if a is a nonisolated essential singularity, for a suitably chosen sequence of values $z_n \to a$, then $f(z)$ can be made to become infinite, or to approach any finite value whatever. *Hint:* Consider the function $G(z) = 1/[f(z) - b]$. Either $G(z)$ has an isolated essential singularity at a, or every deleted neighborhood of a contains a pole or essential singularity. In any of these events, there is a sequence $z_n \to a$ with $G(z_n) \to \infty$.

12. Let the rational function $f(z)$ have a pole of order r_s at a_s, and a zero of order q_t at b_t, where $s = 1, 2, \ldots, n$ and $t = 1, 2, \ldots, m$. If there are no other zeros or poles for finite values of z, show that $f(z) = C(N/D)$, where

$$N = \prod_{t=1}^{m} (z - b_t)^{q_t}, \quad D = \prod_{s=1}^{n} (z - a_s)^{r_s},$$

and C is a constant. *Hint:* The function $(D/N)f(z)$ is rational with no finite singularities, and hence a polynomial. Since it has no zeros, its degree must be zero by the fundamental theorem of algebra.

13. Prove that a rational function is determined up to an additive constant by the principal parts of its Laurent expansions in the neighborhood of each pole in the closed plane. *Hint:* Use Eq. (39).

14. From Eq. (39) deduce that any rational function may be decomposed into partial fractions plus a polynomial. Here the partial fraction expansion has terms of the form of Eq. (36) for each pole at a_i of order r_i. Note that, from Problem 12, in this case the denominator of the rational function will contain a factor $(z - a_i)^{r_i}$ when the rational function is written as the quotient of two polynomials without common factors.

15. A function $f(z)$ has a first-order pole at $z = 2i$ and at $z = -2i$, and no other singularities in the closed plane. Its only zeros are of the first order, and located at $z = 3$ and $z = 1$. If $f(\infty) = 4$, prove that
$$f(z) = 4\frac{z^2 - 4z + 3}{z^2 + 4}.$$
Hint: Use Problem 12.

16. A function $f(z)$ has a first-order pole with residue 5 at $z = 2$, and a second-order pole at the origin with principal part $4/z^2 - 3/z$. It has no other singularities in the closed plane. If $f(\infty) = 0$, prove that
$$f(z) = 2\frac{z^2 + 5z - 4}{z^2(z - 2)}.$$
Hint: Use Eq. (39).

64. The residue theorem

As in Section 62, let the function $f(z)$ be analytic and single-valued in some deleted neighborhood of a. And let its Laurent expansion, valid for $0 < |z - a| < r_2$ be
$$f(z) = \sum_{k=-\infty}^{\infty} A_k(z - a)^k. \tag{40}$$

The coefficient A_{-1} in this expansion is of particular interest. We call it the residue of $f(z)$ at a. This agrees with the definitions for specific types of singularity given in Section 62.

Let C' be any single positive circuit about a made up entirely of points with $0 < |z - a| < r_2$. Then by Eq. (15), we have for the residue:
$$\text{Res}(a) = A_{-1} = \frac{1}{2\pi i}\int_{C'} f(z)\,dz. \tag{41}$$

It follows that, for the circuit C' about a,

$$\int_{C'} f(z)\, dz = 2\pi i \operatorname{Res}(a). \tag{42}$$

By Cauchy's integral theorem, the integral on the left will be zero if $f(z)$ is analytic, or has a removable singularity, at a. But in this event the Laurent expansion of Eq. (40) will reduce to a Taylor's expansion with no negative powers. Hence in particular $A_{-1} = 0$, $\operatorname{Res}(a) = A_{-1} = 0$, and Eq. (42) will still be true.

Next suppose that $f(z)$ is single-valued and analytic within and on the closed contour C, with the exception of a finite number of points a_k within C. Let each of these n points be an isolated singular point as defined in Section 62. Then we may surround each point a_k with a small closed curve C_k so that the curves C_k all taken negatively, together with the curve C taken positively, bound a region in which $f(z)$ is analytic. Then, by the form of Cauchy's integral theorem given in Section 50, we shall have the sum of the integrals of $f(z)$ over these bounding curves equal to zero. And the integral about C will equal the sum of the integrals about C_k, each taken positively. But if the circuits C_k are sufficiently small, each may be used as the C' of Eq. (42), so that each integral about C_k will equal $2\pi i \operatorname{Res}(a_k)$. It follows that

$$\int_C f(z)\, dz = \sum_{k=1}^{n} \int_{C_k} f(z)\, dz = 2\pi i \sum_{k=1}^{n} \operatorname{Res}(a_k). \tag{43}$$

We formulate this result as *Cauchy's residue theorem:*

Let $f(z)$ be analytic on a closed contour C. Inside C, with the exception of a finite number of isolated singular points a_k, let $f(z)$ be single-valued and analytic. Then

$$\int_C f(z)\, dz = 2\pi i \sum \operatorname{Res}(a_k) \tag{44}$$

where $\sum \operatorname{Res}(a_k)$ is the sum of the residues of $f(z)$ at the singular points a_k enclosed by C.

We shall illustrate the use of this theorem in deriving other important general results in the following sections. Its application to the evaluation of real integrals will be treated in Chapter 10.

65. The argument principle

Let $f(z)$ be analytic and $\neq 0$ on C, a contour made up of one or more closed curves which make up the complete boundary of the domain D. With the exception of a finite number of poles in D, let $f(z)$ be single-valued and analytic in D. Then there can be only a finite number of zeros in D.

Suppose that $z = a$ is a zero of order n of $f(z)$. Then, by Eq. (26), we may write

$$f(z) = (z - a)^n h(z), \qquad (45)$$

where $h(z)$ is not zero at a, and is analytic since it has a Taylor's development. Hence, for a suitable choice of branches,

$$\log f(z) = n \log (z - a) + \log h(z). \qquad (46)$$

From this we find that the logarithmic derivative of $f(z)$ is

$$\frac{f'(z)}{f(z)} = \frac{n}{z - a} + \frac{h'(z)}{h(z)}. \qquad (47)$$

The conditions on $h(z)$ make the last term analytic at $z = a$, so that $f'(z)/f(z)$ has a first-order pole at $z = a$ with residue n.

Suppose next that $z = b$ is a pole of order m of $f(z)$. Then, as noted in deriving Eq. (25), we may write

$$(z - b)^m f(z) = H(z) \quad \text{or} \quad f(z) = (z - b)^{-m} H(z) \qquad (48)$$

where $H(z)$ is analytic at b and different from zero there. From this we find that the logarithmic derivative of $f(z)$ is

$$\frac{f'(z)}{f(z)} = -\frac{m}{z - b} + \frac{H'(z)}{H(z)}. \qquad (49)$$

In view of the conditions on $H(z)$, this shows that $f'(z)/f(z)$ has a first-order pole at $z = b$ with residue $-m$.

Let N denote the number of zeros of $f(z)$ in D, counted with their multiplicities. Thus for zeros at a_k with order n_k, $N = \Sigma n_k$.

And let P denote the number of poles of $f(z)$ in D, counted with their multiplicities. Thus for poles at b_k with order m_k, $P = \Sigma m_k$.

Then the logarithmic derivative of $f(z)$, or $f'(z)/f(z)$ is analytic on C and in D with the exception of the points a_k and b_k. At these points it has simple poles with residues n_k at a_k and $-m_k$ at b_k. Hence an application of the residue theorem to $f'(z)/f(z)$ gives

$$\int_C \frac{f'(z)}{f(z)} dz = 2\pi i(N - P). \tag{50}$$

This proves the following theorem:

Let $f(z)$ be analytic and $\neq 0$ on C, the complete boundary of D. Then if $f(z)$ is analytic in D except for a finite number of poles,

$$\frac{1}{2\pi i} \int_C \frac{f'(z)}{f(z)} dz = N - P, \tag{51}$$

which is the number of zeros diminished by the number of poles of $f(z)$ in D, each point counted as often as its order requires.

Corollary. If $f(z)$ is analytic in D, then

$$\frac{1}{2\pi i} \int_C \frac{f'(z)}{f(z)} dz = N. \tag{52}$$

Since there are no poles, $P = 0$ in Eq. (51) and Eq. (52) follows. To bring out the geometric meaning of our results, we write

$$\phi = \arg f(z), \quad R = |f(z)|, \quad f(z) = Re^{i\phi}. \tag{53}$$

Then along any part of C from z_1 to z_2, we have

$$\int_{z_1}^{z_2} \frac{f'(z)}{f(z)} dz = \left[\log f(z)\right]_{z_1}^{z_2} = \text{Log } R_2 - \text{Log } R_1 + i(\phi_2 - \phi_1). \tag{54}$$

For a closed part of C, $R_2 = R_1$. Thus

$$\int_C \frac{f'(z)}{f(z)} dz = i \Delta_C \phi = i \Delta_C \arg f(z), \tag{55}$$

where Δ_C denotes the variation taken over the contour C. This enables us to replace the conclusion of the theorem, Eq. (51), by

$$N - P = \frac{1}{2\pi} \Delta_C \arg f(z). \tag{56}$$

When $f(z)$ is analytic in D, we may replace Eq. (52) by

$$N = \frac{1}{2\pi} \Delta_C \arg f(z). \tag{57}$$

This result, as well as the preceding more general relation, is known as the *argument principle*. We may see its meaning intuitively by considering the mapping of the domain D in the z-plane on to a domain D' in the w-plane. If no value w_0 is taken on more than once in D by $w = f(z)$, then D' will be a simple domain. But, if any value w_0 is taken on at least twice, D' will overlap itself. The number of zeros of $f(z)$ in D is the number of times the origin is covered by D'. If there are no poles of $f(z)$ in D, then D' will be finite and the number of times D' covers the origin is the same as the number of times the boundary C' of D' surrounds the origin. This explains Eq. (57). The number of poles of $f(z)$ in D equals the number of times $w = +\infty$ is covered by D'. But a curve which surrounds the origin once in the negative sense at the same time surrounds the point $w = -\infty$, that is, a domain containing this point, once in the positive sense. Thus each covering of the point $w = +\infty$ will have the effect of diminishing by one unit the number of times C' surrounds the origin. This explains eq. (56). Since the variation over a closed path of $\arg f(z)$ is either zero or a positive or negative integral number of revolutions, the right member of Eq. (56) or Eq. (57) is seen to always be zero or a positive or negative integer. This is less obvious for the left member of Eq. (51) or Eq. (52).

66. Rouché's theorem

We shall now use the argument principle to prove the following result, known as *Rouché's theorem*.

Let each of the functions $f(z)$ and $g(z)$ be single-valued and analytic in a domain D and on its boundary C. Then $|g(z)| < |f(z)|$ on C, the functions $f(z)$ and $f(z) + g(z)$ have the same number of zeros in D.

We observe that the condition $|g(z)| < |f(z)|$ implies that $f(z) \neq 0$ on C, and also $f(z) + g(z) \neq 0$ on C, since on C we have

$$|f(z) + g(z)| > |f(z)| - |g(z)| > 0. \quad (55)$$

Hence we may apply the argument principle, Eq. (57), to show that:

$$N = \frac{1}{2\pi} \Delta \arg f(z), \quad N' = \frac{1}{2\pi} \Delta \arg \{f(z) + g(z)\} \quad (59)$$

where N is the number of zeros of $f(z)$ in D, and N' is the number of zeros of $f(z) + g(z)$. But we have

$$\log(f+g) = \log f + \log\left(1 + \frac{g}{f}\right), \tag{60}$$

and on taking the imaginary part,

$$\arg(f+g) = \arg f + \arg\left(1 + \frac{g}{f}\right). \tag{61}$$

From this and Eq. (59) we may conclude that

$$N' - N = \frac{1}{2\pi}\Delta_C \arg\left(1 + \frac{g}{f}\right). \tag{62}$$

Let $(1 + g/f) = w = Re^{i\phi}$ on C. Since $|g| < |f|$, then $|1 - w| < 1$. Thus on C the point w moves inside the circle $|1 - w| < 1$ which does not contain the origin as an interior point. It follows that the total variation of ϕ on C is zero, or $\Delta_C \phi = \Delta_C \arg(1 + g/f) = 0$. Hence by Eq. (62) $N' = N$ and the theorem is proved.

Exercise 32

1. Verify that if $f(z) = (z - 2)/z(z - 1)$, the residues at the simple poles are Res $(0) = 2$, Res $(1) = -1$. *Hint:* Use Eq. (23) with $m = 1$, or Problem 4 of Exercise 31.

2. Verify that if $f(z) = (z - 2)/z^2(z - 1)^2$, the residues at the double poles are Res $(0) = -3$, Res $(1) = 3$. *Hint:* Expand $z^2 f(z)$ in powers of z, and expand $(z - 1)^2 f(z)$ in powers of $z - 1$. Or use Problem 5 of Exercise 31.

3. Verify that if $f(z) = (z^2 - 9)/(z^2 + 1)$, the residues at the simple poles are Res $(i) = 5i$, Res $(-i) = -5i$. See hint to Problem 1.

Let C_1 be the circle $|z - 1 - i| = 2$ traversed positively.

4. Evaluate $\int_{C_1} f(z)\, dz$ for each of the functions of Problems 1 to 3. *Ans.* $2\pi i$ for Problem 1, 0 for Problem 2, -10π for Problem 3.

5. Evaluate $\int_{C_1} f'(z)/f(z)\, dz$ for each of the functions of Problems 1 to 3. *Ans.* $-2\pi i$ for Problem 1, $-6\pi i$ for Problem 2, $-2\pi i$ for Problem 3.

6. Show that if $f(z)$ has a zero of order n at a, the residue of the function $zf'(z)/f(z)$ at a is na.

7. Show that if $f(z)$ has a pole of order m at b, the residue of the function $zf'(z)/f(z)$ at b is $-mb$.

8. With the notation and hypothesis of the first theorem of Section 65, deduce from Problems 6 and 7 that $\dfrac{1}{2\pi i}\int_C z\dfrac{f'(z)}{f(z)}\,dz = \Sigma\, n_k a_k - \Sigma\, m_k b_k$.

9. Generalize Problems 6 and 7 by showing that if $g(z)$ is any function analytic at a and b, the residue of $g(z)\dfrac{f'(z)}{f(z)}$ at a is $n\,g(a)$, and at b is $-m\,g(b)$.

10. With the notation and hypothesis of the first theorem of Section 65, deduce from Problem 9 that

$$\frac{1}{2\pi i}\int_C g(z)\frac{f'(z)}{f(z)}\,dz = \Sigma\, n_k g(a_k) - \Sigma\, m_k g(b_k).$$

11. Prove that the equation $z^4 + 2z^3 + 3z + 4 = 0$ has exactly one root in the first quadrant. *Hint:* Let $f(z)$ be the left member, and consider $\Delta \arg f$ round the part of the first quadrant bounded by $|z| = R$. On the positive x-axis, f is positive and $\arg f = 0$. On the quadrant of a circle, $\Delta \arg f = \Delta \arg R^4 e^{4i\theta} + \Delta \arg [1 + O(R^{-1})] = 2\pi + O(R^{-1})$. On the axis of y, $\arg f = \tan^{-1}[(-2y^3 + 3y)/(y^4 + 4)]$. The fraction in y is zero at 0, $\sqrt{\tfrac{3}{2}}$, and at $+\infty$. But since the denominator is never zero, the $\arg f$ is always in the first or fourth quadrant, and $\Delta \arg f = 0$ over the positive y-axis from $+\infty$ to 0. Hence for sufficiently large R, $\dfrac{1}{2\pi}\Delta \arg f = N = 1$.

12. Show that all four roots of the equation $z^4 + 6z + 1 = 0$ must lie in the circle $|z| < 2$, but that only one root of this equation lies inside the circle $|z| < \tfrac{2}{3}$.

13. Give another proof of the fundamental theorem of algebra by taking $f(z) = a_n z^n$, $g(z) = a_{n-1} z^{n-1} + \ldots + a_1 z + a_0$ in Rouché's theorem, and using as the contour a circle $|z| = R$ with sufficiently large radius R.

14. If $q > e$, use Rouché's theorem to prove that $e^z = qz^n$ has exactly n roots inside the circle $|z| = 1$.

15. Use continuity considerations along the real axis to show that the one root of $e^z = qz$ with $|z| < 1$ for $q > e$ given by Problem 14 must be real and positive.

16. Show that the function $f(z) = z + 1/z$ assumes every nonreal

value, as well as those real values x such that $x > 2$, or $x < -2$, precisely once inside the unit circle $|z| < 1$. *Hint:* Let z traverse C, the unit circle $z = e^{i\theta}$. Then $w = f(z) = 2\cos\theta$ traverses the segment $-2 \leq x \leq 2$ twice as θ increases from 0 to 2π. Hence for w_0 any value outside this segment, $\Delta_C \arg(w - w_0) = 0$. Hence for the function $f(z) - w_0$, $N - P = 0$. But $P = 1$.

17. Let $f(z)$ be single-valued and analytic on C and in D except for P poles. Prove that if $f(z)$ is real and $\neq 0$ on C, then the number of zeros of $f(z)$ in D is also P.

67. Residues of rational functions

Let the point I or $z = \infty$ be a regular point or an isolated singularity of a function $f(z)$. And let its Laurent expansion, valid for $|z| > 1/r_2$ as in Eq. (32), be

$$f(z) = \sum_{k=-\infty}^{\infty} B_k z^k. \tag{63}$$

A simple circuit which surrounds the point $w = \infty$, that is, a domain containing this point, once in the positive sense, must at the same time surround the origin once in the negative sense. Let C' be such a circuit made up entirely of points with $|z| > 1/r_2$. Then termwise integration of the uniformly convergent series of Eq. (63) about C' shows that

$$\int_{C'} f(z)\,dz = B_{-1} \int_{C'} \frac{dz}{z} = -2\pi i B_{-1}. \tag{64}$$

Thus, to preserve the definition of residue in terms of an integral of Eq. (42), we must define the residue of $f(z)$ at infinity to be

$$\text{Res}(\infty) = -B_{-1} = \frac{1}{2\pi i} \int_{C'} f(z)\,dz. \tag{65}$$

We anticipated this in Section 62. With this definition, we may extend the residue theorem of Section 64 to infinite regions containing only isolated singularities.

In particular, let us again consider a rational function or quotient of two polynomials. We characterized such functions in Section 63 in terms of their singularities, a finite number of poles in the closed plane, including I. Let $f(z)$ be a particular rational function having poles a_k in the finite plane. Let C be a circuit about the origin, with radius so large that it includes all the a_k in its interior. Then by the residue theorem,

$$\frac{1}{2\pi i}\int_C f(z)\,dz = \Sigma \text{ Res }(a_k). \tag{66}$$

But we may consider $-C$ as the C'' of Eq. (65). This shows that

$$-\frac{1}{2\pi i}\int_C f(z)\,dz = \frac{1}{2\pi i}\int_{C'} f(z)\,dz = \text{Res }(\infty). \tag{67}$$

It follows from the last two equations that

$$\Sigma \text{ Res }(a_k) + \text{Res }(\infty) = 0. \tag{68}$$

This proves our first theorem:

The sum of the residues of a rational function at all the poles in the finite plane, together with the residue at infinity, is zero.

For a rational function, the expansion of Eq. (63) will have only a finite number of positive powers, and we may write

$$f(z) = z^K\left(C_0 + \frac{C_1}{z} + \frac{C_2}{z^2} + \ldots\right). \tag{69}$$

It follows that the logarithmic derivative of $f(z)$ is

$$\frac{f'(z)}{f(z)} = \frac{K}{z} + \frac{-C_1 z^{-2} + \ldots}{C_0 + \ldots} = \frac{K}{z} + \frac{D_2}{z^2} + \frac{D_3}{z^3} + \ldots. \tag{70}$$

Thus the conclusions which we drew from Eqs. (47) and (49) also apply to infinity. For, at infinity, the residue of $f'(z)/f(z)$ is $-K$. This is n, 0, or $-m$ according as $K = -n$, 0, or m, that is, according as $f(z)$ has a zero of the nth order, a regular point, or a pole of the mth order at infinity. Thus the theorem which we deduced from Eq. (50) may be applied to a neighborhood of I or $z = \infty$. In particular, consider any closed curve C not passing through any of the zeros or poles of $f(z)$. Then we may consider C taken positively to bound its interior, and taken negatively to bound its exterior. Since these two cancel, it follows that

In the closed plane, any rational function has as many poles as it has zeros.

This number is the *degree* of the rational function. It is equal to the highest power which appears in either numerator or denominator when the rational function is written as the quotient of two polynomials with no common factors. Let the numerator be of degree n and the denominator of degree m. If $m \geq n$, there are m poles in the finite plane, and the function is analytic at infinity. But if $n > m$, there are m poles in the finite plane, and a pole of the $(n - m)$th order at infinity, making $m + (n - m) = n$ poles in the closed plane.

Let $f(z)$ be any rational function of the nth degree. Then it has n poles in the closed plane. If w_0 is any complex number, the function $f(z) - w_0$ will have the same n poles. Hence, by the theorem just proved, it will have n zeros. That is, properly counting multiplicities, $f(z) - w_0 = 0$ and $f(z) = w_0$, n times. This proves our third theorem:

A rational function of degree n assumes any given complex value precisely n times in the closed plane.

68. Conformal mapping of domains

We shall use the argument principle to prove the following property of conformal maps.

Let $f(z)$ be single-valued and analytic in a domain D and on its boundary C, a simple closed path. If the function $f(z)$ maps the curve C in a one-to-one fashion onto a simple closed path C' in the w-plane, then $w = f(z)$ maps D into the domain D' bounded by C'.

We have to prove that, if z is in D then $w = f(z)$ is in D', and that every value of w corresponding to a point in D' is assumed exactly once by $f(z)$ for z a point in D. Let z_0 be a point in D, and let $w_0 = f(z_0)$. Then by the amplitude principle, Eq. (57),

$$N = \frac{1}{2\pi} \Delta_C \arg [f(z) - w_0] \qquad (71)$$

is the number of times w_0 is assumed by $f(z)$ for z in D. Hence the right member must be a positive integer. But the one-to-one mapping of C on C' shows that a single circuit of C must correspond to a positive or negative circuit of C'; and $\Delta_{C'} \arg (w - w_0)$ would be 0 for w_0 outside of D', $\pm \pi$ for w_0 on C', and $\pm 2\pi$ for w_0 inside D'. Only $+2\pi$ makes N a positive integer $+1$, so that w_0 is inside D', and a positive circuit of C corresponds to a positive circuit of C'. This shows that for any w_1 in D', we must have

$$\Delta_C \arg [f(z) - w_1] = \Delta_{C'} \arg (w - w_1) = 2\pi. \qquad (72)$$

Hence every such value is taken on exactly once by $f(z)$ for z in D, as we set out to prove.

Corollary. Under the conditions of the theorem, $f'(z) \neq 0$ for z in D.

To prove this corollary, let z_0 be any point in D. Then if $f'(z_0) = 0$, the function $f(z) - f(z_0)$ would have at least a double zero at $z = z_0$. Thus the value $f(z_0)$ would be assumed more than once by $f(z)$ for z in D, contrary to the conclusion of the theorem.

Without giving the proof, we note that there is an extension of the theorem of this section which asserts that $w = f(z)$ maps the domain D bounded by C into the domain D' bounded by C' provided that $f(z)$ is analytic in D, and continuous on C, and maps the properly oriented circuit C into the properly oriented circuit C'. This modified result also applies to regions extending to infinity.

69. The Schwarz-Christoffel transformation

Because of its many applications, we shall briefly describe a general expression for a function $w(z)$ which will map the upper half of the z-plane on the interior of a given polygon P in the w-plane.

Let the interior angles of the polygon be β_k, so that if there are n vertices $\Sigma \beta_k = (n - 2)\pi$. And let x_k be n points on the real axis of the z-plane such that $x_1 < x_2 < \ldots < x_n$.

Then the transformation $w(z)$ may be defined by

$$w(z) = G \int (z - x_1)^{p_1}(z - x_2)^{p_2} \ldots (z - x_n)^{p_n} dz + H, \qquad (73)$$

or by the relation

$$\frac{dw}{dz} = G(z - x_1)^{p_1}(z - x_2)^{p_2} \ldots (z - x_n)^{p_n}, \qquad (74)$$

where the exponents p_k are defined by

$$p_k = \frac{\beta_k}{\pi} - 1. \qquad (75)$$

For any choice of the complex constants G and H, the function $w(z)$ maps the real axis in the z-plane onto some polygon in the w-plane having the given interior angle β_k at the vertex $w(x_k)$. This follows from Eq. (74), which shows that $\arg dw/dz$ is constant along each segment $x_{k-1} < x < x_k$. But as x increases through x_k, $z - x_k$ rotates through $-\pi$, so that $\arg dw/dz$ increases by $\pi - \beta_k$, the appropriate exterior angle of the polygon. It may be shown that if any three of the x_k are chosen arbitrarily, the $(n-3)$ remaining x_k and the two complex constants G, H may be so determined that the function $w(z)$ maps the real axis onto the given polygon P.

The form of dw/dz given in Eq. (74) shows that all its singularities are on the real axis and that dw/dz is never zero in the upper half z-plane. Hence $w(z)$ is analytic in this upper half-plane.

That $w(z)$ must map this upper half-plane on the interior of the polygon P then follows from the modified result which was stated at the end of Section 68.

For P a finite polygon of n vertices, we may make the nth vertex correspond to the point at infinity on the x-axis. By incorporating a factor $(-x_n)^{-p_n}$ in G, and then letting $x_n \to \infty$, so that $(z - x_n)^{p_n}(-x_n)^{-p_n} \to 1$, this will have the effect of removing the last factor in Eqs. (74) and (73) which will then have only $(n-1)$ factors.

The transformation may be applied in some cases where the polygon P does not have all its vertices located in the finite part of the plane, as will be seen from Problems 4 to 6 of Exercise 33.

Exercise 33

1. Use the fact that there is an nth-order pole at infinity, and the second theorem of Section 67, to prove again that a polynomial of the nth degree has n roots. Note that this also follows directly from the third theorem of Section 67.

2. Check the second theorem of Section 67 by noting that the quotient of a polynomial of the nth degree by one of the mth degree, prime to it, has the n roots of the numerator as zeros, the m roots of the denominator as poles, and infinity as a zero of order $m - n$, an

ordinary point, or a pole of order $n - m$, according as $m > n$, $m = n$, or $m < n$.

3. For the function of Problem 1 of Exercise 32, verify directly that $\text{Res}(\infty) = -1$. Note that the function is analytic, and in fact has a first-order zero at $z = \infty$. Thus $N = P = 2$, and $\text{Res}(0) + \text{Res}(1) + \text{Res}(\infty) = 2 - 1 - 1 = 0$, as it should.

In each of the following mappings, the upper half of the z-plane is mapped on a certain degenerate polygon of the w-plane, which is described at the beginning of the problem. Verify the indicated correspondence of certain boundary points, the expression for dw/dz, and the relation of the exponents to interior angles of Eq. (75).

4. Infinite strip $(-\infty < u < \infty, 0 \leqq v \leqq b)$.

Mapping: $z = e^{\pi w/b}$. $\quad \dfrac{dw}{dz} = \dfrac{b}{\pi z}$. $\quad$ Correspondences: $u = +\infty$, $z = I = \infty$; $w = bi$, $z = -1$; $u = -\infty$, $w = 0$; $w = 0$, $z = 1$.

5. Semi-infinite strip $(0 \leqq u \leqq a, 0 \leqq v < \infty)$.
Mapping:
$$z = -\cos \frac{\pi w}{a}. \qquad \frac{dw}{dz} = \frac{a}{\pi} \frac{1}{\sqrt{1-z^2}}.$$

Correspondences:
$$v = +\infty, \quad z = I = \infty; \qquad w = 0, \quad z = -1;$$
$$w = \frac{a}{2}, \quad z = 0; \qquad w = a, \quad z = 1.$$

6. Infinite sector $(0 < \arg w < \alpha, 0 \leqq |w| < \infty)$.
Mapping:
$$z = w^{\pi/\alpha}. \qquad \frac{dw}{dz} = \frac{\alpha}{\pi} z^{\alpha/\pi - 1}.$$

Correspondences:
$$w = (Re^{i\alpha})_{R \to \infty}, \quad z = -\infty;$$
$$w = 0, \quad z = 0; \qquad u = +\infty, \quad z = +\infty.$$

10

APPLICATIONS OF RESIDUES

This chapter begins with a recapitulation of some definitions and results pertaining to the residues of a function at its isolated singularities. We then explain how a number of types of real definite integrals may be evaluated by means of Cauchy's residue theorem. We also use the theorem to obtain some expansions in rational fractions and the sums of certain infinite series.

70. Isolated singular points

Let the function $f(z)$ be analytic and single-valued in some deleted neighborhood of a. And let its Laurent expansion, valid for $0 < |z - a| < r_2$ be

$$f(z) = \sum_{k=-\infty}^{\infty} A_k(z - a)^k. \tag{1}$$

Then the *residue* of $f(z)$ at a, abbreviated as Res (a), is the coefficient of $(z - a)^{-1}$ in the Laurent expansion, or A_{-1}.

Let C' be any single positive circuit about a made up entirely of points with $0 < |z - a| < r_2$. Then, as shown in Section 64, or from Eq. (1) by termwise integration,

$$\text{Res}(a) = A_{-1} = \frac{1}{2\pi i} \int_{C'} f(z)\, dz. \tag{2}$$

It follows that, for the circuit C' about a,

$$\int_{C'} f(z)\, dz = 2\pi i\, \text{Res}(a). \tag{3}$$

If $f(z)$ is analytic or has a removable singularity at a, this equation remains true. For in either case there are then no negative powers in the expansion of Eq. (1), so that $A_{-1} = 0$, and the left member of Eq. (3) is zero by Cauchy's integral theorem. Usually after a singularity is known to be removable, it is reduced to the analytic case by defining or redefining the function at a with the appropriate value. Thus for the most part such values are not included among the singularities. But the remark made above shows that if we do apply Eq. (3) or its consequences to a value a which is a removable singularity, no harm is done.

When there are an infinite number of negative powers in the expansion of Eq. (1), we have an *essential singularity*. When there are only a finite number of negative powers, we have a *pole*. The pole is of *order* m if $(z - a)^{-m}$ is the highest power of $(z - a)^{-1}$ in the expansion.

As we showed in detail in Section 64, an application of a relation like Eq. (3) to each singular point a_k, combined with Cauchy's integral theorem may be used to establish *Cauchy's residue theorem*:

Let $f(z)$ be analytic on a closed contour C. Inside C, with the exception of a finite number of isolated singular points a_k, let $f(z)$ be single-valued and analytic. Then

$$\int_C f(z)\, dz = 2\pi i\, \Sigma\, \text{Res}(a_k), \tag{4}$$

where $\Sigma\, \text{Res}(a_k)$ is the sum of the residues of $f(z)$ at the points a_k enclosed by C.

71. Residue at infinity

Let the point I or $z = \infty$ be a regular point or an isolated singularity of a function $f(z)$, as in Section 67. And let its Laurent expansion, valid for $|z| > 1/r_2$, be

$$f(z) = \sum_{k=-\infty}^{\infty} B_k z^k. \tag{5}$$

Then the residue of $f(z)$ at infinity is the negative of the coefficient of z^{-1}, or $-B_{-1}$ and

$$\text{Res}(\infty) = -B_{-1} = \frac{1}{2\pi i} \int_{C'} f(z)\, dz, \tag{6}$$

where C' is a simple circuit which surrounds I once in the positive sense, and therefore at the same time surrounds the origin once in the negative sense, and is made up entirely of points with $|z| > 1/r_2$. With this definition we may extend the residue theorem to infinite regions containing only isolated singularities, as was illustrated in Section 67. For such regions, we must always include I among the a_k, even though $f(z)$ be regular at I. For a function may be regular at I, and still have a nonzero residue there. For example, the function $1/z$ has a residue -1 at I or $z = \infty$. But it is regular at I, since if $f(z) = 1/z$, $f(1/Z) = Z$ is regular at $Z = 0$.

72. Evaluation of residues

For a few very simple functions with an essential singularity, the Laurent expansion may be found by transforming a Taylor's series in $Z = (z - a)^{-1}$, and this leads to the residue. Thus for $z \cosh(3/z)$, the residue at $z = 0$ is $\frac{3}{2}$, since

$$z \cosh \frac{3}{z} = z\left(1 + \frac{1}{2!}\frac{3^2}{z^2} + \frac{1}{4!}\frac{3^4}{z^4} + \ldots\right). \tag{7}$$

If $f(z)$ has a simple pole at $z = a$, the Laurent expansion is

$$f(z) = \frac{A_{-1}}{z - a} + A_0 + A_1(z - a) + A_2(z - a)^2 + \ldots, \tag{8}$$

with $A_{-1} \neq 0$. Hence in this case

$$\lim_{z \to a} (z - a)f(z) = A_{-1}. \tag{9}$$

Conversely, whenever the limit on the left is finite and not zero, $f(z)$ has a simple pole at a, with Res $(a) =$ this limit. In particular, let

$$f(z) = \frac{P_0 + P_1(z-a) + P_2(z-a)^2 + \ldots}{Q_1(z-a) + Q_2(z-a)^2 + \ldots} = \frac{p(z)}{q(z)}, \quad (10)$$

with $P_0 \neq 0$ and $Q_1 \neq 0$. Then in this case, from Eq. (9) or by division, we find that

$$\text{Res}(a) = \frac{P_0}{Q_1} = \frac{p(a)}{q'(a)} \quad (11)$$

Next let $f(z)$ have a pole of order m at $z = a$. Then

$$f(z) = \frac{A_{-m}}{(z-a)^m} + \frac{A_{-m+1}}{(z-a)^{m-1}} + \ldots + \frac{A_{-1}}{z-a}$$
$$+ A_0 + A_1(z-a) + A_2(z-a)^2 + \ldots. \quad (12)$$

And in place of Eq. (9) we have

$$\lim_{z \to a}(z-a)^m f(z) = A_{-m}. \quad (13)$$

Conversely, if the limit on the left is finite and not zero, then $f(z)$ has a pole of order m at a. But if $m > 1$, this does not give the residue. However, from the discussion of Section 62,

$$G(z) = (z-a)^m f(z) \quad \text{if } z \neq a, \text{ and } G(a) = A_{-m} \quad (14)$$

is analytic at a. And from Eq. (12), its Taylor's series is

$$G(z) = A_{-m} + \ldots + A_{-1}(z-a)^{m-1} + A_0(z-a)^m + \ldots. \quad (15)$$

It follows that

$$\text{Res}(a) = A_{-1} = \frac{1}{(m-1)!}\left[\frac{d^{m-1}G(z)}{dz^{m-1}}\right]_{z=a}. \quad (16)$$

In particular, let

$$f(z) = \frac{P_0 + P_1(z-a) + P_2(z-a)^2 + \ldots}{Q_m(z-a)^m + Q_{m+1}(z-a)^{m+1} + \ldots} = \frac{p(z)}{q(z)}, \quad (17)$$

with $P_0 \neq 0$ and $Q_m \neq 0$. Then in this case

$$G(z) = \frac{P_0 + P_1(z-a) + P_2(z-a)^2 + \ldots}{Q_m + Q_{m+1}(z-a) + Q_{m+2}(z-a)^2 + \ldots}. \quad (18)$$

And the residue of $f(z)$ at a is the coefficient of $(z-a)^{m-1}$ in the expansion of this fraction, which may sometimes be found by division more easily than by using Eq. (16).

For example, if $m = 2$ in Eqs. (17) and (18),

$$A_{-1} = \frac{P_1 Q_2 - P_0 Q_3}{Q_2^2} = \frac{6p'(a)q''(a) - 2p(a)q'''(a)}{3[q''(a)]^2} \quad (19)$$

Sec. 73 APPLICATIONS OF RESIDUES 213

Exercise 34

For each of the following functions, verify the statements made about its residues at its singularities and at infinity.

1. $\dfrac{z+2}{z^2-3z}$. Res $(0) = -\tfrac{2}{3}$, Res $(3) = \tfrac{5}{3}$, Res $(\infty) = -1$. Simple poles at 0 and 3. Regular point at ∞.

2. $\dfrac{z^4}{z^2+1}$. Res $(i) = -i/2$, Res $(-i) = i/2$, Res $(\infty) = 0$. Simple poles at i and $-i$. A pole of order 2 at ∞.

3. $\dfrac{1}{z^3(z-1)^2}$. Res $(0) = 3$, Res $(1) = -3$, Res $(\infty) = 0$. A pole of order 3 at 0, of order 2 at 1. Regular point at ∞.

Let C be the circle $\left|z - \dfrac{i}{2}\right| = 1$ traversed positively. Show that

4. $\displaystyle\int_C \dfrac{z+2}{z^2-3z}\,dz = -\dfrac{4\pi i}{3}$, by using the result of Problem 1.

5. $\displaystyle\int_C \dfrac{z^4}{z^2+1}\,dz = \pi$, by using the result of Problem 2.

6. $\displaystyle\int_C \dfrac{dz}{z^3(z-1)^2} = 6\pi i$, by using the result of Problem 3.

7. Verify that the function $z/(\sin z)$ has a removable singularity at the origin, and a simple pole at $z = n\pi$ with Res $(n\pi) = (-1)^n n\pi$ for n equal to a positive integer or to a negative integer. Note that the essential singularity at infinity is not isolated, so that a residue at infinity cannot be defined.

8. Verify that for $z^2 e^{1/z}$, Res $(0) = \tfrac{1}{6}$, Res $(\infty) = -\tfrac{1}{6}$. Essential singularity at 0, pole of order 2 at ∞. Deduce that if C is the circle $|z| = 2$ traverses positively, $\displaystyle\int_C z^2 e^{1/z}\,dz = \pi i/3$.

9. Verify that for $e^{1/z}/(z-1)^2$, Res $(0) = e$, Res $(1) = -e$, Res $(\infty) = 0$. Essential singularity at 0, pole of order 2 at 1, regular point at ∞.

10. Verify that for $(\cosh 2z)/z^5$, Res $(0) = \tfrac{2}{3}$, Res $(\infty) = -\tfrac{2}{3}$. Pole of order 3 at 0, essential singularity at ∞.

73. Real integrals found by using the unit circle

Let $\phi(\cos\theta, \sin\theta)$ be the quotient of two polynomials in $\cos\theta$ and sin θ, each with real coefficients. And let the quotient never become infinite for any real value of θ. Then the real integral

$$I = \int_0^{2\pi} \phi(\cos\theta, \sin\theta)\, d\theta \qquad (20)$$

may be found by putting $z = e^{i\theta}$, so that

$$\cos\theta = \frac{z + z^{-1}}{2}, \quad \sin\theta = \frac{z - z^{-1}}{2i}, \quad d\theta = \frac{dz}{iz} \qquad (21)$$

This reduces I to the integral of a rational function about the unit circle C, with radius one and center at the origin. Since there are no poles on this circle, the integral may be evaluated in terms of the residues at the poles inside the unit circle C.

Example. Prove that for $a > b > 0$,

$$I = \int_0^{\pi} \frac{\sin^4\theta\, d\theta}{a + b\cos\theta} = \frac{\pi}{b^4}\left[-a^3 + \tfrac{3}{2}ab^2 + (a^2 - b^2)^{3/2}\right]. \qquad (22)$$

Solution: Since $\sin^4\theta$ is an even function, and $\sin\theta$ and $\cos\theta$ each have the period 2π,

$$\int_0^{\pi} = \tfrac{1}{2}\int_{-\pi}^{\pi} = \tfrac{1}{2}\int_0^{2\pi}.$$

Hence from Eq. (21) we have

$$I = \frac{1}{2}\int_C \frac{(z - z^{-1})^4}{(2i)^4}\frac{2}{2a + b(z + z^{-1})}\frac{dz}{iz}$$

$$= \frac{1}{16bi}\int_C \frac{(z^2 - 1)^4\, dz}{z^4(z^2 + 2az/b + 1)} = \frac{1}{16bi}\int_C f(z)\, dz. \qquad (23)$$

The integrand has poles at 0, r_1, and r_2, where r_1, r_2 are the roots of $z^2 + 2az/b + 1 = 0$, or

$$r_1 = \frac{-a + \sqrt{a^2 - b^2}}{b}, \qquad r_2 = \frac{-a - \sqrt{a^2 - b^2}}{b} \qquad (24)$$

Since $a > b > 0$, then $|r_2| > a/b > 1$, and since $r_1 r_2 = 1$, then $|r_1| < 1$. Hence for points inside C, 0 is a pole of the fourth order and r_1 is a simple pole for the function

$$f(z) = \frac{(z^2 - 1)^4}{z^4(z - r_1)(z - r_2)}. \qquad (25)$$

At r_1, we have

APPLICATIONS OF RESIDUES

$$\text{Res}(r_1) = \lim_{z \to r_1}(z - r_1)f(z) = \frac{(r_1^2 - 1)^4}{r_1^4(r_1 - r_2)}$$

$$= \frac{(r_1 - r_1^{-1})^4}{(r_1 - r_2)} = (r_1 - r_2)^3 = \frac{2^3(a^2 - b^2)^{3/2}}{b^3}.$$

To find the residue of $f(z)$ at 0, we need the coefficient of z^3 in the expansion of

$$\frac{(z^2 - 1)^4}{(z - r_1)(z - r_2)} = \frac{1 - 4z^2 + 6z^4 - \cdots}{1 + 2az/b + z^2}$$

$$= 1 - \frac{2az}{b} + \left(\frac{4a^2}{b^2} - 5\right)z^2 + \left(-\frac{8a^3}{b^3} + \frac{12a}{b}\right)z^3 + \cdots.$$

It then follows from Eq. (25) that $\text{Res}(0) = -8a^3/b^3 + 12a/b$.

We may now deduce from Eq. (23) that

$$I = \frac{2\pi i}{16bi}[\text{Res}(r_1) + \text{Res}(0)] = \frac{\pi}{b^4}[(a^2 - b^2)^{3/2} - a^3 + \tfrac{3}{2}ab^2]. \quad (26)$$

74. Infinite integrals of rational functions

Let the function $Q(z)$ be a rational function of z with real coefficients. Let the degree of the numerator be n, and that of the denominator be $n + s$, with $s \geq 2$. Then if the denominator has no real zeros, the integral $\int_{-\infty}^{\infty} Q(x)\,dx$ may be found by the use of residues. Call the zeros of the denominator, or poles of $Q(z)$, in the *upper* half-plane $a_1, a_2, \ldots, a_p$.

Let S_M denote a semicircle in the upper half-plane with center at the origin and radius M, traversed in the positive direction from M to $-M$. Our contour consists of the part of the real axis from $-M$ to M, together with the semicircle S_M traversed in the positive direction. We choose M so large that this contour includes in its interior all the zeros of the denominator which lie in the upper half-plane, $a_1, a_2, \ldots, a_p$. Then from Eq. (4) we have

$$\int_{-M}^{M} Q(z)\,dz + \int_{S_M} Q(z)\,dz = 2\pi i \sum_{k=1}^{p} \text{Res}(a_k). \quad (27)$$

Our conditions on the degree make

$$\lim_{z \to \infty} |z^s Q(z)| = K' < K, \quad |Q(z)| < K|z|^{-s} \quad (28)$$

for $|z|$ sufficiently large. It follows that $\int_{-\infty}^{\infty} Q(x)\, dx$ converges, and that

$$\int_{-\infty}^{\infty} Q(x)\, dx = \lim_{M \to \infty} \int_{-M}^{M} Q(z)\, dz. \tag{29}$$

Also, for M sufficiently large, on S_M we have $|Q(z)| < KM^{-2}$,

$$\left| \int_{S_M} Q(z)\, dz \right| < \pi M K M^{-2}, \quad \text{and} \quad \lim_{M \to \infty} \int_{S_M} Q(z)\, dz = 0. \tag{30}$$

It now follows from Eqs. (27), (29), and (30), that

$$\int_{-\infty}^{\infty} Q(x)\, dx = 2\pi i \sum_{k=1}^{p} \operatorname{Res}(a_k). \tag{31}$$

This gives the evaluation of the integral provided that the denominator of the rational function $Q(z)$ has no real zeros, has zeros $a_1, a_2, \ldots, a_p$ in the upper half plane, and is of degree at least *two* higher than that of the numerator.

We note that if $Q(z)$ is an even function,

$$\int_0^{\infty} Q(x)\, dx = \tfrac{1}{2} \int_{-\infty}^{\infty} Q(x)\, dx. \tag{32}$$

Example. Prove that for $b > 0$,

$$\int_{-\infty}^{\infty} \frac{dx}{x^2 + 2ax + a^2 + b^2} = \frac{\pi}{b}. \tag{33}$$

Solution: The zeros of the denominator are $r_1 = -a + bi$, $r_2 = -a - bi$.

At the simple pole r_1 the residue of the function $(z - r_1)^{-1}(z - r_2)^{-1}$ may be found from Eq. (9) as $(r_1 - r_2)^{-1} = (2bi)^{-1}$, or from Eq. (11) with $p(z) = 1$, $q(z) = z^2 + 2az + a^2 + b^2$, $q'(z) = 2z + 2a$ as $(2r_1 + 2a)^{-1} = (2bi)^{-1}$. From this, the value of the integral in Eq. (33) is $2\pi i \operatorname{Res}(r_1) = 2\pi i / 2bi = \pi/b$.

Exercise 35

Use the method of residues to verify that

1. $\int_0^{\pi} \dfrac{d\theta}{a + b \cos \theta} = \dfrac{\pi}{\sqrt{a^2 - b^2}}, a > |b|$.

2. $\int_0^{2\pi} \frac{d\theta}{a + b \sin\theta} = \frac{2\pi}{\sqrt{a^2 - b^2}}, a > |b|.$

3. $\int_0^{\pi} \frac{d\theta}{a + \sin^2\theta} = \frac{\pi}{\sqrt{a^2 + a}}, a > 0.$

4. $\int_0^{\pi} \frac{\sin^2\theta \, d\theta}{a + b\cos\theta} = \frac{\pi}{b^2}(a - \sqrt{a^2 - b^2}), a > |b| > 0.$

5. $\int_0^{\pi/2} \sin^4\theta \, d\theta = \frac{1}{4}\int_0^{2\pi} \sin^4\theta \, d\theta = \frac{3\pi}{16}.$

6. $\int_0^{\pi/2} \cos^{2n}\theta \, d\theta = \frac{\pi}{2}\frac{1 \cdot 3 \cdot 5 \ldots (2n-1)}{2 \cdot 4 \cdot 6 \ldots 2n}$, n a positive integer.

From the indicated problem, by differentiation with respect to a, deduce that

7. $\int_0^{\pi} \frac{d\theta}{(a + b\cos\theta)^2} = \frac{\pi a}{(a^2 - b^2)^{3/2}}, a > |b|$, from Problem 1.

8. $\int_0^{\pi} \frac{d\theta}{(a + \sin^2\theta)^2} = \frac{\pi(2a + 1)}{2(a^2 + a)^{3/2}}, a > 0$, from Problem 3.

Use the method of residues to verify that for $a > 0$

9. $\int_0^{\infty} \frac{dx}{(x^2 + a^2)(x^2 + b^2)} = \frac{\pi}{2ab(a + b)}, a > 0, b > 0.$

10. $\int_0^{\infty} \frac{dx}{x^4 + a^4} = \frac{\pi\sqrt{2}}{4a^3}.$ 11. $\int_0^{\infty} \frac{dx}{x^6 + a^6} = \frac{\pi}{3a^5}.$

12. $\int_0^{\infty} \frac{x^2 \, dx}{x^6 + a^6} = \frac{\pi}{6a^3}.$ 13. $\int_0^{\infty} \frac{dx}{(x^2 + a^2)^2} = \frac{\pi}{4a^3}.$

14. $\int_0^{\infty} \frac{dx}{x^4 - 6x^2 + 25} = \frac{\pi}{20}.$ Hint: $x^4 - 6x^2 + 25 = (x^2 + 5)^2 - (4x)^2.$

15. From Eq. (33), by differentiating n times with respect to b, deduce that

$$\int_{-\infty}^{\infty} \frac{dx}{[(x + a)^2 + b^2]^n} = \frac{\pi 1 \cdot 3 \cdot 5 \ldots (2n - 3)}{b^{2n-1}2^{n-1}(n-1)!}$$
$$= \frac{\pi(2n - 2)!}{b^{2n-1}2^{2n-2}[(n-1)!]^2}.$$

16. Verify that for the function
$$\frac{z^{-n}}{(1+p^2)z - p(z^2+1)}, \quad p < 1,$$
Res $(0) = \dfrac{p^n - p^{-n}}{1 - p^2}$. *Hint:* The function is
$$\frac{z^{-n}}{-p(z-p)(z-p^{-1})} = \frac{z^{-n}}{1-p^2}\left(\frac{1}{z-p} - \frac{1}{z-p^{-1}}\right).$$
And the parenthesis is $-\Sigma\, p^{-k-1}z^k + \Sigma\, p^{k+1}z^k$, k from 0 to ∞, with $p^n - p^{-n}$ as the coefficient of z^{n-1}.

17. From Problem 16, deduce that for $p < 1$,
$$\int_0^\pi \frac{\cos n\theta\, d\theta}{1 - 2p\cos\theta + p^2} = \frac{\pi p^n}{1 - p^2}, \quad n \text{ a positive integer}.$$

By the method used for Problems 16, 17 or by putting $\theta = 2\theta'$ in Problem 17, deduce that for $p < 1$, n a positive integer,

18. $\displaystyle\int_0^\pi \frac{\cos^2 n\theta\, d\theta}{1 - 2p\cos 2\theta + p^2} = \frac{\pi}{2}\frac{1 + p^n}{1 - p^2}.$

19. $\displaystyle\int_0^\pi \frac{\sin^2 n\theta\, d\theta}{1 - 2p\cos 2\theta + p^2} = \frac{\pi}{2}\frac{1 - p^n}{1 - p^2}.$

75. Infinite integrals with sin mx or cos mx as a factor

For some rational functions $Q(z)$, the integrals
$$\int_{-\infty}^{\infty} Q(x)\sin mx\, dx \quad \text{and} \quad \int_{-\infty}^{\infty} Q(x)\cos mx\, dx$$
may be found by the use of residues.

Suppose that $Q(z)$ is a rational function whose denominator is of degree at least one higher than that of the numerator. Then for M sufficiently large, $|Q(z)| < KM^{-1}$ on S_M, the semicircle of Section 74. On S_M we have
$$z = Me^{i\theta} = M\cos\theta + iM\sin\theta, \quad dz = iMe^{i\theta}\, d\theta. \tag{34}$$
Consequently, for sufficiently large M, on S_M
$$|e^{imz}Q(z)| < KM^{-1}e^{-mM\sin\theta}. \tag{35}$$

For $0 < \theta < \pi/2$, $\theta < \tan\theta$ and $\sin\theta/\theta$ has a negative derivative.

Hence as θ increases from 0 to $\pi/2$, $\sin\theta/\theta$ decreases from 1 to $2/\pi$. Thus $\sin\theta/\theta > 2/\pi$ for $0 < \theta < \pi/2$, and

$$\sin\theta \geq \frac{2\theta}{\pi} \quad \text{for} \quad 0 \leq \theta \leq \frac{\pi}{2}. \tag{36}$$

Let us now assume that $m > 0$. Then

$$\int_0^\pi e^{-mM\sin\theta}\,d\theta = 2\int_0^{\pi/2} e^{-mM\sin\theta}\,d\theta \leq 2\int_0^{\pi/2} e^{-2mM\theta/\pi}\,d\theta$$

$$\leq \left[\frac{-\pi}{mM} e^{-2mM\theta/\pi}\right]_0^{\pi/2} < \frac{\pi}{mM}. \tag{37}$$

It follows from Eqs. (34), (35), and (37) that

$$\left|\int_{S_M} e^{imz}Q(z)\,dz\right| \leq \int_0^\pi KM^{-1}e^{-mM\sin\theta}M\,d\theta < \frac{K\pi}{mM}. \tag{38}$$

Since the right member tends to 0 as $M \to \infty$, this proves that

$$\lim_{M \to \infty} \int_{S_M} e^{imz}Q(z)\,dz = 0. \tag{39}$$

We assume that the denominator of $Q(z)$ has no real zeros. Call the zeros of the denominator in the *upper* half-plane $a_1, a_2, \ldots, a_p$. As in Section 74, we consider the contour consisting of the part of the real axis from $-M$ to M together with the semicircle S_M. We choose M so large that this contour includes in its interior all the poles $a_1, a_2, \ldots, a_p$ of $Q(z)$ which lie in the upper half-plane. Then from the residue theorem, Eq. (4), applied to the function $e^{imz}Q(z)$ with m any *positive* number, we have

$$\int_{-M}^M e^{imz}Q(z)\,dz + \int_{S_M} e^{imz}Q(z)\,dz = 2\pi i \sum_{k=1}^p \text{Res}\,(a_k). \tag{40}$$

And it then follows from this and Eq. (39) that

$$\lim_{M \to \infty} \int_{-M}^M e^{imz}Q(z)\,dz = 2\pi i \sum_{k=1}^p \text{Res}\,(a_k). \tag{41}$$

We note that

$$\int_{-M}^M e^{imz}Q(z)\,dz = \int_0^M e^{imz}Q(z)\,dz + \int_0^M e^{-imz}Q(-z)\,dz, \tag{42}$$

where we have replaced z by $-z$ in the integral from $-M$ to 0.

Hence, if $Q(z)$ is an *even* function of z, $Q(-z) = Q(z)$ and

APPLICATIONS OF RESIDUES

$$\lim_{M \to \infty} \int_{-M}^{M} e^{imz} Q(z)\, dz = 2 \int_{0}^{\infty} Q(z) \cos mz\, dz. \qquad (43)$$

Again, if $Q(z)$ is an *odd* function of z, then $Q(-z) = -Q(z)$ and

$$\lim_{M \to \infty} \int_{-M}^{M} e^{imz} Q(z)\, dz = 2i \int_{0}^{\infty} Q(z) \sin mz\, dz. \qquad (44)$$

Thus for $Q(z)$ a rational function with real coefficients, whose denominator is at least one degree higher than its numerator, with the denominator having no real roots and roots $a_1, a_2, \ldots, a_p$ in the upper half-plane, we have the following results.

For $Q(z)$ an *even* function:

$$\int_{0}^{\infty} Q(x) \cos mx\, dx = \pi i \sum_{k=1}^{p} \text{Res}(a_k). \qquad (45)$$

For $Q(z)$ an *odd* function:

$$\int_{0}^{\infty} Q(x) \sin mx\, dx = \pi \sum_{k=1}^{p} \text{Res}(a_k). \qquad (46)$$

In Eqs. (45) and (46), $m > 0$ and the residues on the right are those of the function $e^{imz} Q(z)$.

For $Q(z)$ neither even nor odd, we may take real and imaginary parts in Eq. (41) and so evaluate $\int_{-\infty}^{\infty} Q(x) \cos mx\, dx$ as well as $\int_{-\infty}^{\infty} Q(x) \sin mx\, dx$. These necessarily converge because of our condition on $Q(z)$, as may be shown by decomposing $Q(z)$ into partial fractions, taking real and imaginary parts, and applying Abel's test for convergence.

Exercise 36

Use the method of residues to verify that for $m > 0$, $a > b > 0$,

1. $\int_{0}^{\infty} \dfrac{\cos mx}{a^2 + x^2}\, dx = \dfrac{\pi e^{-am}}{2a}$.

2. $\int_{0}^{\infty} \dfrac{x \sin mx}{a^2 + x^2}\, dx = \dfrac{\pi e^{-am}}{2}$. Check by Problem 1 and differentiation with respect to m.

3. $\int_{0}^{\infty} \dfrac{\cos mx\, dx}{(x^2 + a^2)(x^2 + b^2)} = \dfrac{\pi}{2(a^2 - b^2)} \left(\dfrac{e^{-bm}}{b} - \dfrac{e^{-am}}{a} \right)$. Check by using partial fractions and Problem 1.

4. $\int_0^\infty \dfrac{x \sin mx\, dx}{(x^2+a^2)(x^2+b^2)} = \dfrac{\pi}{2(a^2-b^2)}(e^{-bm}-e^{-am})$. Check by using partial fractions and Problem 2.

5. $\int_0^\infty \dfrac{\cos mx}{x^4+4a^4}\, dx = \dfrac{\pi}{8a^3} e^{-am}(\cos am + \sin am)$.

6. $\int_0^\infty \dfrac{\cos mx\, dx}{(x^2+a^2)^2} = \dfrac{\pi}{4a^3} e^{-am}(am+1)$. Check by Problem 1 and differentiation with respect to a, or from Problem 3 as $b \to a$.

7. $\int_0^\infty \dfrac{x \sin mx}{(x^2+a^2)^2}\, dx = \dfrac{\pi m}{4a} e^{-am}$. Check by Problem 2 and differentiation with respect to a, or from Problem 4 as $b \to a$.

8. $\int_0^\infty \dfrac{x^2 \cos mx}{(x^2+a^2)^2}\, dx = \dfrac{\pi}{4a} e^{-am}(1-am)$. Check by Problem 7 and differentiation with respect to m.

9. $\int_{-\infty}^\infty \dfrac{\cos mx\, dx}{(x+c)^2+a^2} = \dfrac{\pi}{a} e^{-am} \cos cm$, and
$\int_{-\infty}^\infty \dfrac{\sin mx\, dx}{(x+c)^2+a^2} = -\dfrac{\pi}{a} e^{-am} \sin cm$, for c real.

10. $\int_0^\infty \dfrac{x \sin mx}{x^4+4a^4}\, dx = \dfrac{\pi}{4a^2} e^{-am} \sin am$. Check by Problem 5 and differentiation with respect to m.

11. $\int_0^\infty \dfrac{x^2 \cos mx}{x^4+4a^4}\, dx = \dfrac{\pi}{4a} e^{-am}(\cos am - \sin am)$. Check by Problem 10 and differentiation with respect to m.

12. $\int_0^\infty \dfrac{x^3 \sin mx}{x^4+4a^4}\, dx = \dfrac{\pi}{2} e^{-am} \cos am$. Check by Problem 11 and differentiation with respect to m.

76. Principal value of an integral

The integral of a rational function over an interval including one of its real roots diverges. For example $\int_0^3 \dfrac{dx}{x-2}$ diverges, since the integrand becomes infinite at $x=2$, and as $x \to 2$, the indefinite integral $\log|x-2| \to -\infty$.

Suppose that we omit an interval with center at the singularity, in this case from $2-h$ to $2+h$, and let $h \to 0$. For any small positive h we have

$$\int_0^{2-h} \frac{dx}{x-2} = \left[\log|x-2|\right]_0^{2-h} = \log h - \log 2,$$

$$\int_{2+h}^8 \frac{dx}{x-2} = \left[\log|x-2|\right]_{2+h}^8 = \log 6 - \log h.$$

Thus the sum of the integrals is log 3, which approaches the limit log 3 when we let $h \to 0$. The number obtained by this process is known as the *principal value* of the integral, and is indicated by P, so that

$$P \int_0^8 \frac{dx}{x-2} = \lim_{h \to 0} \left[\int_0^{2-h} \frac{dx}{x-2} + \int_{2+h}^8 \frac{dx}{x-2}\right] = \log 3. \quad (47)$$

The term *Cauchy principal value* is sometimes used, and the symbol P is often omitted in technical literature.

Similarly, the integral $\int_{-\infty}^{\infty} \frac{2x\,dx}{x^2+1}$ diverges, since the indefinite integral $\log(x^2+1)$ becomes infinite at infinity. But since the integrand is odd, we have

$$P \int_{-\infty}^{\infty} \frac{2x\,dx}{x^2+1} = \lim_{M \to \infty} \left[\int_{-M}^{M} \frac{2x\,dx}{x^2+1}\right] = 0. \quad (48)$$

77. Indented contours

Let us now consider the integral $P \int_{-\infty}^{\infty} Q(x)\,dx$, where the function $Q(z)$ is a rational function of z with real coefficients. As in Section 74, we again require that the degree of the denominator of $Q(z)$ be at least two higher than that of the numerator. But we now permit this denominator to have simple real zeros. Thus in addition to the complex zeros $a_1, a_2, \ldots, a_p$ in the upper half-plane, there may be simple real zeros $b_1, b_2, \ldots, b_q$. Any one such zero causes the integral to diverge, but the principal value defined in Section 76 exists.

We first consider the case of a single real zero at b. Then we construct a circle of radius m and center at b, taking m so small that $Q(z)$ is analytic for $0 < |z-b| \leq m$. Let s denote the upper half of this circle traversed *negatively*, from $b-m$ to $b+m$. Then we modify the contour of Section 74 by replacing the straight-line segment from

$b - m$ to $b + m$ by the semicircle s. Then for M sufficiently large, we have in place of Eq. (27),

$$\int_{-M}^{b-m} Q(z)\,dz + \int_s Q(z)\,dz + \int_{b+m}^{M} Q(z)\,dz + \int_{S_M} Q(z)\,dz$$
$$= 2\pi i \sum_{k=1}^{p} \text{Res}\,(a_k). \quad (49)$$

Whenever a contour includes part of a small circle with center b (which might be complex in some cases) to avoid a singularity of the integrand $f(z)$ at b, the contour is said to be *indented* at b. Let the indented arc A_{12} from z_1 to z_2 be that part of the circle included between the lines $\text{Arg}\,(z - b) = \theta_1$ and $\text{Arg}\,(z - b) = \theta_2$. Then if $f(z)$ has a simple pole at b, the limit of the integral of $f(z)$ along A_{12} when the radius of the circle $m \to 0$ may be found as follows.

For any point z on A_{12} let $\theta = \text{Arg}\,(z - b)$. Then

$$z - b = me^{i\theta} \quad \text{and} \quad \log(z - b) = \log m + i\theta. \quad (50)$$

Since $\text{Arg}\,(z_1 - b) = \theta_1$, $\text{Arg}\,(z_2 - b) = \theta_2$, we have

$$\int_{A_{12}} \frac{dz}{z - b} = \left[\log(z - b)\right]_{z_1}^{z_2} = \left[\log m + i\theta\right]_{\theta_1}^{\theta_2} = i(\theta_2 - \theta_1). \quad (51)$$

For a simple pole, we may deduce from Eq. (8) that

$$f(z) = \frac{\text{Res}\,(b)}{z - b} + \phi(z), \quad (52)$$

where $\text{Res}\,(b)$ is the residue of $f(z)$ at b, and $\phi(z)$ is analytic at b. Along A_{12}, $z \to b$ when $m \to 0$. Hence M, the maximum value of $|\phi(z)|$ on A_{12}, approaches $|\phi(b)|$. But L, the length of A_{12}, approaches zero. Hence by the inequality of Section 46, the integral of $\phi(z)$ along A_{12} approaches zero when $m \to 0$. Consequently we have

$$\lim_{m \to 0} \int_{A_{12}} f(z)\,dz = \lim_{m \to 0} \int_{A_{12}} \frac{\text{Res}\,(b)}{z - b}\,dz = i(\theta_2 - \theta_1)\,\text{Res}\,(b). \quad (53)$$

If A_{12} is C_b, a complete positive circuit about b, $\theta_1 = 0$, $\theta_2 = 2\pi$, $\theta_2 - \theta_1 = 2\pi$, and the factor $2\pi i$ is in accord with the residue theorem. If A_{12} is s, a semicircle traversed in the negative direction, $\theta_1 = \pi$, $\theta_2 = 0$, $\theta_2 - \theta_1 = -\pi$ and

$$\lim_{m \to 0} \int_s f(z)\,dz = -\pi i\,\text{Res}\,(b). \quad (54)$$

224   APPLICATIONS OF RESIDUES   Chap. 10

We may now let $M \to \infty$ and $m \to 0$ in Eq. (49). Then the sum of the first and third integrals approaches $P\int_{-\infty}^{\infty} Q(z)\, dz$. The fourth integral approaches zero, as was shown in Section 74. And the limit of the second integral is found from Eq. (54), so that

$$P\int_{-\infty}^{\infty} Q(z)\, dz - \pi i \operatorname{Res}(b) = 2\pi i \sum_{k=1}^{p} \operatorname{Res}(a_k). \qquad (55)$$

When there are q simple poles on the real axis, we indent the contour at each of the points b_k by a semicircle s_k of radius m_k from $b_k - m_k$ to $b_k + m_k$. And by reasoning similar to that just used, we obtain a relation like Eq. (55) with $\sum \operatorname{Res}(b_k)$ in place of $\operatorname{Res}(b)$. Thus we find

$$P\int_{-\infty}^{\infty} Q(z)\, dz = 2\pi i \sum_{k=1}^{p} \operatorname{Res}(a_k) + \pi i \sum_{k=1}^{q} \operatorname{Res}(b_k). \qquad (56)$$

This is the modification of Eq. (31) to be used when $Q(z)$ has simple poles $b_1, b_2, \ldots, b_q$ on the real axis.

The corresponding modification of Eq. (41) is

$$P\int_{-\infty}^{\infty} e^{imz}Q(z)\, dz = 2\pi i \sum_{k=1}^{p} \operatorname{Res}(a_k) + \pi i \sum_{k=1}^{q} \operatorname{Res}(b_k). \qquad (57)$$

As a simple example of this, let $Q(z) = 1/(z - b)$. Then for the function $e^{imz}/(z - b)$, at the real pole b, $\operatorname{Res}(b) = e^{imb}$. Hence

$$P\int_{-\infty}^{\infty} \frac{e^{imz}}{z - b}\, dz = \pi i e^{imb},$$

so that

$$P\int_{-\infty}^{\infty} \frac{\cos mx}{x - b}\, dx = -\pi \sin mb, \quad P\int_{-\infty}^{\infty} \frac{\sin mx}{x - b}\, dx = \pi \cos mb. \qquad (58)$$

Exercise 37

1. Let $Q_1(z)$ be a rational function whose denominator is of degree exactly one more than that of its numerator. Then its Laurent expansion of Eq. (5) is

$$Q_1(z) = \frac{B_{-1}}{z} + \frac{B_{-2}}{z^2} + \ldots = \frac{B_{-1}}{z} + Q_2(z). \qquad (59)$$

Sec. 77 APPLICATIONS OF RESIDUES 225

Show that $\lim_{z \to \infty} zQ_1(z) = B_{-1} = -\text{Res}(\infty)$, and that $\lim_{z \to \infty} z^2 Q_2(z) = B_{-2}$. From this and Eq. (59), by reasoning as we did for Eqs. (30) and (51), deduce that

$$\lim_{M \to \infty} \int_{S_M} Q_1(z)\, dz = \pi i B_{-1} = -\pi i \,\text{Res}(\infty). \tag{60}$$

2. Let the function $Q_1(z)$ of Problem 1 have poles a_k in the upper half-plane, and simple real poles b_k on the real axis. From Problem 1, Eq. (60), and Eq. (49), deduce that we now have in place of Eq. (56),

$$P\int_{-\infty}^{\infty} Q_1(z)\, dz = 2\pi i \sum_{k=1}^{p} \text{Res}(a_k)$$
$$+ \pi i \sum_{k=1}^{q} \text{Res}(b_k) + \pi i \,\text{Res}(\infty). \tag{61}$$

3. As applications of Problem 2, verify that for $a > 0$,

$$P\int_{-\infty}^{\infty} \frac{dx}{x-b} = 0 \quad \text{and} \quad P\int_{-\infty}^{\infty} \frac{x^2\,dx}{(x^2+a^2)(x-b)} = \frac{\pi ab}{2(a^2+b^2)}.$$

Use the method of residues to verify that for $a > 0$, $m > 0$,

4. $\int_0^\infty \frac{\sin cx}{x}\, dx = \frac{\pi}{2}$ for $c > 0$, and hence $= -\frac{\pi}{2}$ for $c < 0$. The integral obviously $= 0$ for $c = 0$.

5. $\int_0^\infty \frac{\sin mx}{x(a^2+x^2)}\, dx = \frac{\pi}{2a^2}(1-e^{-am})$. Check by using partial fractions, Problem 2 of Exercise 36, and Problem 4.

6. $\int_0^\infty \frac{\sin^2 mx}{x^2}\, dx = \frac{\pi m}{2}$. Hint: $\sin^2 mx = \text{Re}\left(\frac{1-e^{2imx}}{2}\right)$. Check by using integration by parts and Problem 4.

7. $\int_0^\infty \frac{\sin^2 mx\, dx}{x^2(a^2+x^2)} = \frac{\pi}{4a^4}(e^{-2am} - 1 + 2am)$. See hint to Problem 6.

8. $\int_0^\infty \frac{mx - \sin mx}{x^3(a^2+x^2)}\, dx = \frac{\pi}{2a^4}\left(-e^{-am} + 1 - am + \frac{a^2 m^2}{2}\right).$

9. From Problem 5, by differentiation with respect to a, deduce that

$$\int_0^\infty \frac{\sin mx\, dx}{x(a^2+x^2)^2} = \frac{\pi}{2a^4}\left(1 - \frac{2+am}{2} e^{-am}\right).$$

226　APPLICATIONS OF RESIDUES　Chap. 10

10. From Problem 5, by differentiation with respect to m, check the result of Problem 1 of Exercise 36.

11. From Problem 8, by differentiation with respect to m twice, check Problem 5. And by putting $m = 2m'$ in the first derivative, check Problem 7.

12. Prove that

$$P\int_0^\pi \frac{\cos n\theta \, d\theta}{\cos \theta - \cos \phi} = \pi \frac{\sin n\phi}{\sin \phi},$$

for n zero or a positive integer. This integral is used in airfoil theory. *Hint:* Let C be the unit circle, indented at $e^{i\phi}$ and $e^{-i\phi}$. And let I be the given integral. Then

$$2I - \pi i \operatorname{Res}(e^{i\phi}) - \pi i \operatorname{Res}(e^{-i\phi})$$
$$= \int_C \frac{(z^n + z^{-n})\, dz}{i(z - e^{i\phi})(z - e^{-i\phi})} = 2\pi i \operatorname{Res}(0).$$

Then

$$\operatorname{Res}(e^{i\phi}) = \frac{-\cos n\phi}{\sin \phi}, \quad \operatorname{Res}(e^{-i\phi}) = \frac{\cos n\phi}{\sin \phi}.$$

At 0 we may write the integrand as

$$\frac{z^n + z^{-n}}{-2 \sin \phi}\left(\frac{1}{z - e^{i\phi}} - \frac{1}{z - e^{-i\phi}}\right).$$

The parentheses is $-\sum e^{-ik\phi}z^{k-1} + \sum e^{ik\phi}z^{k-1}$, k from 0 to ∞, with $e^{in\phi} - e^{-in\phi} = 2i \sin n\phi$ as the coefficient of z^{n-1}. Hence $\operatorname{Res}(0) = \frac{1}{i}\frac{\sin n\phi}{\sin \phi}$, for $n > 0$, and $\operatorname{Res}(0) = 0$ for $n = 0$.

13. From Problem 12, with $n = 0$, deduce that

$$P\int_0^\pi \frac{\cos n\phi \, d\theta}{\cos \theta - \cos \phi} = 0,$$

and therefore the convergent integral

$$\int_0^\pi \frac{\cos n\theta - \cos n\phi}{\cos \theta - \cos \phi} d\theta = \pi \frac{\sin n\phi}{\sin \phi}.$$

78. Integrals with a many-valued factor

For some rational functions $Q(z)$, the integral $\int_0^\infty x^{p-1} Q(x)\, dx$ with $0 < p < 1$ may be found by the use of residues. We assume that

$Q(z)$ has no poles on the *positive* real axis. We also require that $z^p Q(z)$ tends to zero uniformly both as z tends to zero, and as z tends to infinity. We consider the integral $\int_C z^{p-1} Q(z) \, dz$, where C is a contour made up of a large circle C_M of radius M, a small circle c_m of radius m, each with center at the origin, and a cut along the positive real axis traversed twice. Specifically, the contour proceeds from m to M on the "upper edge," or upper bank of the cut along the real axis, then around the large circle C_M in the positive directon to M on the "lower edge," or lower bank of the cut, from M to m on the lower bank of the cut along the real axis, and around c_m in the negative direction to the starting point m on the upper bank. The choice of this contour is motivated by the fact that, as discussed in Section 29, z^p is not single-valued in the z-plane, or in any complete neighborhood of its branch point at zero or of that at infinity. But z^p is single-valued in the domain bounded by C, since this domain is contained in the z-plane which has been cut along the entire positive real axis. Hence we may apply Cauchy's residue theorem to deduce that

$$\left(\int_m^M U + \int_{C_M} + \int_M^m L - \int_{c_m} \right) z^{p-1} Q(z) \, dz = 2\pi i \, \Sigma \, \text{Res} \, (a_k). \quad (62)$$

Here U denotes upper and L denotes lower bank, and the summation is over all the poles of $Q(z)$ in the domain bounded by C. This will be all the poles of $Q(z)$ when M is sufficiently large, and m is sufficiently small.

For $|z^p Q(z)| < \epsilon_1$ on c_m, we have the inequality of Section 46,

$$\left| \int_{c_m} z^{p-1} Q(z) \, dz \right| < \epsilon_1 \frac{1}{m} 2\pi m \quad \text{or} \quad 2\pi \epsilon_1. \quad (63)$$

It follows that as $m \to 0$, the $\int_{c_m} \to 0$ in Eq. (62).

Similarly for $|z^p Q(z)| < \epsilon_2$ on C_M, we have

$$\left| \int_{C_M} z^{p-1} Q(z) \, dz \right| < \epsilon_2 \frac{1}{M} 2\pi M \quad \text{or} \quad 2\pi \epsilon_2. \quad (64)$$

It follows that as $M \to \infty$, the $\int_{C_M} \to 0$ in Eq. (62).

For a value of $z = x$ on the upper bank of the cut along the positive real axis, let us take $\arg z = 0$ and $z^p = x^p$. Then

$$\int_m^M (U) z^{p-1} Q(z)\, dz = \int_m^M x^{p-1} Q(x)\, dx \to \int_0^\infty x^{p-1} Q(x)\, dx \quad (65)$$

as $m \to 0$, $M \to \infty$.

When z traverses C_M in the positive direction, $\arg z$ increases by $2\pi i$. Hence for $z = x$ on the lower bank of the cut along the positive real axis, $z = xe^{2\pi i}$ and $z^p = x^p e^{2\pi i p}$ so that

$$\int_M^m (L) z^{p-1} Q(z)\, dz = \int_M^m x^{p-1} e^{2\pi i p} Q(x)\, dx \to -e^{2\pi i p} \int_0^\infty x^{p-1} Q(x)\, dx. \quad (66)$$

We may now let $m \to 0$ and $M \to \infty$ in Eq. (62) to obtain

$$\int_0^\infty x^{p-1} Q(x)\, dx - e^{2\pi i p} \int_0^\infty x^{p-1} Q(x)\, dx = 2\pi i \sum \operatorname{Res}(a_k). \quad (67)$$

This leads to the evaluation

$$\int_0^\infty x^{p-1} Q(x)\, dx = \frac{2\pi i}{1 - e^{2\pi i p}} \sum \operatorname{Res}(a_k). \quad (68)$$

The residues on the right are those of the function $z^{p-1} Q(z)$. If $a_k = r e^{i\varphi}$, we must take $z^p = r^p e^{i\varphi p}$ so that

$$\operatorname{Res}(a_k) = \frac{r^p e^{i\varphi p}}{a_k} \quad [\text{Residue of } Q(z) \text{ at } a_k]. \quad (69)$$

Example. Prove that

$$\int_0^\infty \frac{x^{p-1}}{1+x}\, dx = \frac{\pi}{\sin \pi p} \quad \text{for} \quad 0 < p < 1. \quad (70)$$

Solution: Here $Q(z) = 1/(1+z)$, $z^p Q(z) = z^p/(1+z) \to 0$ uniformly as $z \to 0$ or as $z \to \infty$. Thus Eq. (68) is applicable, with the single pole at $z = -1 = 1 e^{\pi i}$. At -1, $z^p = e^{\pi i p}$, and

$$\operatorname{Res}(-1) = \lim_{z \to -1} (z+1) \frac{z^{p-1}}{1+z} = -e^{\pi i p}.$$

Thus from Eq. (68),

$$\int_0^\infty \frac{x^{p-1}}{1+x}\, dx = \frac{2\pi i(-e^{\pi i p})}{1 - e^{2\pi i p}} = \frac{\pi(2i)}{e^{\pi i p} - e^{-\pi i p}} = \frac{\pi}{\sin \pi p}. \quad (71)$$

79. Special types of contour

So far, for the most part we have used circular and semicircular contours or slight modifications of these. But for some purposes other types of contour, such as sectors or rectangles, may be more effective. Several applications of special contours either to the direct evaluation of integrals, or to the reduction of new integrals to known ones, are included in the problems of Exercise 38.

Exercise 38

1. Verify that $\int_0^\infty \frac{x^q\,dx}{a^2 + x^2} = \frac{\pi a^{q-1}}{2\cos(\pi q/2)}$ for $-1 < q < 1, a > 0$. Check by putting $x = u^2/a^2$, $p = (q+1)/2$ in Eq. (71).

2. Verify that $\int_0^\infty \frac{x^q\,dq}{(a^2 + x^2)^2} = \frac{\pi a^{q-3}(1-q)}{4\cos(\pi q/2)}$ for $-1 < q < 3$, $a > 0$. Check for $-1 < q < 1$ by using Problem 1 and differentiation with respect to a.

3. By putting $x = u^{2n}/a^{2n}$, $p = (2m+1)/2n$ in Eq. (71), deduce that
$$\int_0^\infty \frac{x^{2m}\,dx}{x^{2n} + a^{2n}} = \frac{\pi a^{2m-2n+1}}{2n \sin[(2m+1)\pi/2n]}$$
for $a > 0$ and all real m, n with $0 < 2m + 1 < 2n$, hence in particular for m zero or an integer and n an integer such that $0 \le m < n$. Use this to check Problems 10, 11, and 12 of Exercise 35.

4. Verify that
$$\int_0^\infty \frac{x^q\,dx}{a^2 + 2ax\cos\theta + x^2} = \frac{\pi a^{q-1}}{\sin \pi q} \cdot \frac{\sin q\theta}{\sin \theta}$$
for $-1 < q < 1$, $-\pi < \theta < \pi$, $a > 0$. *Hint:* For the roots of the denominator, to keep $\arg a_1$ and $\arg a_2$ in the proper range, we may take $a_1 = e^{i(\pi+\theta)}$ and $a_2 = e^{i(\pi-\theta)}$. The result checks Problem 1 when $\theta = \pi/2$.

5. By integrating $z^{p-1}/(1-z)$ about a contour consisting of a large semicircle in the upper half-plane S_M, together with the real axis indented by a semicircle at $z = 0$, and a semicircle at $z = 1$, deduce that for $0 < p < 1$,
$$\int_{-\infty}^0 \frac{z^{p-1}}{1-z}\,dz + P\int_0^\infty \frac{z^{p-1}}{1-z}\,dz = -\pi i.$$

Let
$$I_1 = \int_0^\infty \frac{x^{p-1}}{1+x}\,dx \quad \text{and} \quad I_2 = P\int_0^\infty \frac{x^{p-1}}{1-x}\,dx.$$

Then $-e^{\pi i p}I_1 + I_2 = -\pi i$. Taking imaginary parts gives $I_1 = \pi/(\sin \pi p)$, as in Eq. (71). Then taking real parts gives $-I_1 \cos \pi p + I_2 = 0$, so that

$$P\int_0^\infty \frac{x^{p-1}}{1-x}\,dx = \pi \cot \pi p, \qquad 0 < p < 1.$$

6. From Problem 5, with $p = 1 - p'$, deduce the convergent integral

$$\int_0^\infty \frac{x^{-p} - x^{-q}}{1-x}\,dx = \pi(\cot \pi q - \cot \pi p), \qquad 0 < p < 1, \ 0 < q < 1.$$

7. In the theory of the gamma function, it is shown that for $p > 0$, $q > 0$,

$$B(p,q) = \int_0^1 x^{p-1}(1-x)^{q-1}\,dx = \frac{\Gamma(p)\Gamma(q)}{\Gamma(p+q)}.$$

Since $\Gamma(1) = 1$,

$$\int_0^1 x^{p-1}(1-x)^{-p}\,dx = \Gamma(p)\Gamma(1-p)$$

if $0 < p < 1$. Put $x = t/(1+t) = 1 - 1/(1+t)$ in the integral to reduce it to $\int_0^\infty \frac{t^{p-1}}{1+t}\,dt$. Then deduce from Eq. (71) that

$$\Gamma(p)\Gamma(1-p) = \frac{\pi}{\sin \pi p}.$$

8. For $p > 0$, the gamma function

$$\Gamma(p) = \int_0^\infty t^{p-1}e^{-t}\,dt.$$

With $t = x^2$, this is $2\int_0^\infty x^{2p-1}e^{-x^2}\,dx$. With $p = \frac{1}{2}$ in this and Problem 7, deduce that

$$\int_0^\infty e^{-x^2}\,dx = \tfrac{1}{2}\Gamma(\tfrac{1}{2}) = \frac{\sqrt{\pi}}{2}.$$

9. Integrate e^{-z^2} around a rectangle whose vertices are $-M$, M, $M + ib$, $-M + ib$, where $b > 0$. By letting $M \to \infty$, and using the result of Problem 8, deduce that

$$\int_0^\infty e^{-x^2}\cos 2bx\,dx = \frac{\sqrt{\pi}}{2}e^{-b^2}.$$

10. Integrate $e^{iz}z^{p-1}$ about a contour consisting of the segment

from m to M along the real axis, a quadrant of a circle of radius M to iM, then along the imaginary axis from iM to im, and then along a quadrant of a small circle of radius m from im to m. Assume $0 < p < 1$. Let $m \to 0$ and $M \to \infty$ and reason as in Section 75 for the large quadrant. Hence show that

$$\int_0^\infty e^{ix} x^{p-1}\, dx + \int_\infty^0 e^{-y} y^{p-1} e^{(p-1)i\pi/2} i\, dy = 0.$$

By equating real and imaginary parts, and using the definition of $\Gamma(p)$ given in Problem 8, deduce that for $0 < p < 1$,

$$\int_0^\infty x^{p-1} \cos x\, dx = \Gamma(p) \cos \frac{\pi p}{2},$$

$$\int_0^\infty x^{p-1} \sin x\, dx = \Gamma(p) \sin \frac{\pi p}{2}.$$

11. Integrate e^{-z^2} around the boundary of a sector of angle $\pi/4$, that is, from 0 to M along the real axis, along an arc of a circle of radius M from M to $(1 + i)M/\sqrt{2}$, and then from this point to zero along the line $y = x$. Let $M = \infty$ and reason as in Section 75 for the large arc. With $z = (1 + i)t/\sqrt{2}$ on $y = x$, show that

$$\int_0^\infty e^{-x^2}\, dx + \int_\infty^0 e^{-it^2} \frac{1 + i}{\sqrt{2}}\, dt = 0.$$

By taking real and imaginary parts, and using Problem 8, deduce that

$$\int_0^\infty \cos x^2\, dx = \frac{\sqrt{2\pi}}{4}, \qquad \int_0^\infty \sin x^2\, dx = \frac{\sqrt{2\pi}}{4}.$$

With a change of variable, $x = \sqrt{\pi/2}\, u$, these become

$$\int_0^\infty \cos \frac{\pi u^2}{2}\, du = \tfrac{1}{2}, \qquad \int_0^\infty \sin \frac{\pi u^2}{2}\, du = \tfrac{1}{2},$$

the special values of Fresnel's integrals which give the limit point on Cornu's spiral.

12. Integrate $e^{pz}/(1 + e^z)$ around the rectangle with vertices at $-M$, M, $M + 2\pi i$, $-M + 2\pi i$. Assume $0 < p < 1$. Let $M \to \infty$. Hence show that

$$\int_{-\infty}^\infty \frac{e^{px}}{1 + e^x}\, dx = \frac{\pi}{\sin \pi p}, \quad \text{for } 0 < p < 1.$$

Check by putting $x = e^u$ in Eq. (71). Note that the contour used here is in essence the image under $w = \log z$ of the contour used to derive Eq. (71).

13. Integrate $e^{px}/(1 - e^z)$ around the contour of Problem 12, indented by small semicircles at 0 and $2\pi i$ which are poles of the function. Hence deduce that if $0 < p < 1$,

$$P \int_{-\infty}^{\infty} \frac{e^{px}}{1 - e^x} dx = \pi \cot \pi p.$$

Check by putting $x = e^u$ in Problem 5.

14. From Problem 13, deduce that the convergent integral

$$\int_{-\infty}^{\infty} \frac{e^{px} - e^{qx}}{1 - e^x} dx = \pi(\cot \pi p - \cot \pi q), \quad 0 < p < 1, 0 < q < 1.$$

15. Show that

$$\int_{-\infty}^{\infty} \frac{e^{2iax}}{\cosh \pi x} dx = \sec a, \quad -\frac{\pi}{2} < a < \frac{\pi}{2}.$$

Hint: Put $x = u/2\pi$ in the integral and use Problem 12.

80. Expansions in rational fractions

Let $f(z)$ be analytic at all finite points of the plane, except at the points $a_1, a_2, a_3, \ldots$, where

$$0 \leq |a_1| \leq |a_2| \leq |a_3| \leq \ldots. \tag{72}$$

Let each point a_k be a simple pole of $f(z)$, and let b_k be the residue of $f(z)$ at a_k. We consider any contour C_n, with no poles on the contour, bounding a region including the first m_n poles. From the residue theorem, it follows that

$$\int_{C_n} f(w) \left[\frac{1}{w - z} - \frac{1}{w} \right] dw$$
$$= 2\pi i \left\{ f(z) - f(0) + \sum_{n=1}^{m_n} b_n \left[\frac{1}{a_n - z} - \frac{1}{a_n} \right] \right\}, \tag{73}$$

for any value of z distinct from all the a_k.

If a sequence of contours can be found, such that as $n \to \infty$, the limit of the integral on the left is zero, and $m_n \to \infty$, it follows that

$$f(z) = f(0) + \lim_{m_n \to \infty} \sum_{n=1}^{m_n} b_n \left[\frac{1}{z - a_n} + \frac{1}{a_n} \right]. \tag{74}$$

Sec. 80   APPLICATIONS OF RESIDUES   233

In particular, suppose that the contours are similar figures. Then their lengths and maximum and minimum distances from the origin will all be fixed constants times some dimension M_n. Thus if $|f(w)| \leq K_n|w|$ on C_n, we shall have

$$\left| \int_{C_n} f(w) \left[\frac{1}{w-z} - \frac{1}{w} \right] dw \right| = \left| \int_{C_n} \frac{f(w)z}{w(w-z)} dw \right| \leq K_n |z| q, \quad (75)$$

where q is a constant depending on the shape of the contours, provided that the minimum distance from the origin exceeds $2|z|$, or $|w - z| > |w|/2$. Thus for a fixed z the integral will approach zero if we can select constants K_n approaching zero as $n \to \infty$. Furthermore, for z in any region inside some large circle and outside a number of small circles about the poles a_k contained in the large circle, the integral will approach zero uniformly.

It follows that, under these conditions, the limit in Eq. (74) will be approached uniformly for z in such a region.

In some cases, we may select contours C_n on which $|f(w)|$ is uniformly bounded. We may then take $K_n = q'/M_n$, since this makes $K_n \to 0$ as n and hence $M_n \to \infty$.

As an example, consider the function defined by

$$f(z) = \csc z - \frac{1}{z}, \quad \text{for} \quad z \neq 0, \quad f(0) = 0, \quad (76)$$

so that $f(0)$ is the limit of $f(z)$ at the removable singularity at $z = 0$. Then for $z = x + iy$,

$$|\csc z| = \left| \frac{2i}{e^{ix-y} - e^{-ix+y}} \right| \leq \frac{2}{|e^y - e^{-y}|}$$

$$\leq \frac{2}{e - e^{-1}} \quad \text{if } |y| \geq 1. \quad (77)$$

As the contour C_n, let us take a square with center at the origin, sides parallel to the x and y axes, and passing through the point $(n + \frac{1}{2})\pi$. Then the relation of Eq. (77) gives an upper bound for $|f(w)|$ for the part of the contour with $|y| \geq 1$. For the part with $|y| \leq 1$, we observe that $|\csc z|$ has the period π. Hence we may use an upper bound for $|\csc(\pi/2 + iy)|$ for y in this range. Unity is such a bound, since the reciprocal is

$$\sin\left(\frac{\pi}{2} + iy\right) = \cos iy = \cosh y \geq 1. \tag{78}$$

Thus the expansion of Eq. (74) is valid for the function defined in Eq. (76). The poles are the points $n\pi$, with n a positive or negative integer. At $n\pi$, we have

$$\lim_{z \to n\pi} (z - n\pi) \csc z = (-1)^n \lim_{h \to 0} \frac{h}{\sin h} = (-1)^n. \tag{79}$$

Hence the residue at $n\pi$ is $(-1)^n$. Since each contour brings in two additional poles, we have

$$\csc z - \frac{1}{z} = \lim_{m \to \infty} \sum_{n=-m}^{m}{}' (-1)^n \left[\frac{1}{z - n\pi} + \frac{1}{n\pi}\right]. \tag{80}$$

The prime on the summation sign means that the term for $n = 0$ is to be omitted. For the individual terms of the sum,

$$\left|\frac{1}{z - n\pi} + \frac{1}{n\pi}\right| = \left|\frac{z}{n\pi(z - n\pi)}\right|$$

$$\leq \frac{2|z|}{n^2\pi^2}, \quad \text{if } |n\pi| \geq 2|z|. \tag{81}$$

Consequently, in any finite region not including any of the poles as interior or boundary points, the infinite series corresponding to Eq. (80) converges absolutely. For z in such a region,

$$\csc z = \frac{1}{z} + \sum_{n=1}^{\infty} (-1)^n \frac{2z}{z^2 - n^2\pi^2} \tag{82}$$

$$= \frac{1}{z} + \sum_{n=-\infty}^{\infty}{}' (-1)^n \left[\frac{1}{z - n\pi} + \frac{1}{n\pi}\right]. \tag{83}$$

The first expression, Eq. (82), follows directly from Eq. (80) by combining the two terms for n, a positive integer, and $-n$. And the second expression, Eq. (83), with the terms taken in any order that includes each one at some stage, is equal to the first because of the absolute convergence.

Some similar expansions, and some of their consequences, will be found in the problems of Exercise 39. In particular see Problems 9 and 10 for the product representations of the sine and cosine.

Exercise 39

By reasoning as in the text, show that in any finite region not including any of the poles,

1. $\cot z = \dfrac{1}{z} + \sum\limits_{n=1}^{\infty} \dfrac{2z}{z^2 - n^2\pi^2}$

   $= \dfrac{1}{z} + \sum\limits_{n=-\infty}^{\infty}{}' \left[\dfrac{1}{z - n\pi} + \dfrac{1}{n\pi} \right].$

2. $\tan z = \sum\limits_{n=1}^{\infty} \dfrac{2z}{(2n-1)^2\pi^2/4 - z^2}$

   $= \sum\limits_{n=-\infty}^{\infty} \left[\dfrac{4}{(2n-1)\pi} - \dfrac{1}{z - (2n-1)\pi/2} \right].$

3. $\sec z = \sum\limits_{n=1}^{\infty} (-1)^n \dfrac{(2n-1)\pi}{z^2 - (2n-1)^2\pi^2/4}$

   $= \sum\limits_{n=-\infty}^{\infty} (-1)^n \left[\dfrac{1}{z - (2n-1)\pi/2} + \dfrac{2}{(2n-1)\pi} \right].$

4. $\dfrac{\sin az}{\sin z} = \sum\limits_{n=1}^{\infty} (-1)^n \dfrac{2n\pi \sin n\pi a}{z^2 - n^2\pi^2}, \; -1 < a < 1.$

5. $\dfrac{\cos az}{\sin z} = \dfrac{1}{z} + \sum\limits_{n=1}^{\infty} (-1)^n \dfrac{2z \cos n\pi a}{z^2 - n^2\pi^2}, \; -1 < a < 1.$

6. Check Problem 2 by using Eqs. (82), (83), Problem 1, and the identity $\tan z = \csc 2z - \cot 2z$.

Use the fact that a uniformly convergent series of analytic functions may be differentiated termwise to deduce that:

7. $\csc^2 z = \sum\limits_{n=-\infty}^{\infty} \dfrac{1}{(z - n\pi)^2}$, from Problem 1.

8. $\sec^2 z = \sum\limits_{n=-\infty}^{\infty} \dfrac{1}{[z - (2n-1)(\pi/2)]^2}$, from Problem 2. Check by using $\sec z = \csc(\pi/2 - z)$ and Problem 7.

9. Use Problem 1 and termwise integration in the plane cut by the real axis to show that

$$\int_0^z \left(\cot z - \frac{1}{z}\right) dz = \text{Log} \frac{\sin z}{z} = \sum_{n=-\infty}^{\infty} {}' \int_0^z \left(\frac{1}{z - n\pi} + \frac{1}{n\pi}\right) dz.$$

From this deduce the *product expansion* of the sine, for all z.

$$\sin z = z \prod_{n=-\infty}^{\infty} {}' \left(1 - \frac{z}{n\pi}\right) e^{z/n\pi}$$

$$= z \prod_{n=1}^{\infty} \left(1 - \frac{z^2}{n^2 \pi^2}\right).$$

10. Use Problem 2 and a procedure like that of Problem 9 to deduce that for all z,

$$\cos z = \prod_{n=1}^{\infty} \left(1 - \frac{4z^2}{(2n-1)^2 \pi^2}\right).$$

Check by using the identity $\cos z = \sin(z + \pi/2)$, the first expression of Problem 9, and regrouping the terms.

11. Find the coefficient of z^2 for each member of the second equation of Problem 9, and so deduce that

$$\frac{\pi^2}{6} = \frac{1}{1^2} + \frac{1}{2^2} + \frac{1}{3^2} + \cdots.$$

12. Find the coefficient of z^2 for each member of the equation of Problem 10, and so deduce that

$$\frac{\pi^2}{8} = \frac{1}{1^2} + \frac{1}{3^2} + \frac{1}{5^2} + \cdots.$$

81. Summation of series

For some rational functions $f(z)$, the infinite series $\Sigma f(n)$ may be summed by integrating $\pi (\cot \pi z) f(z)$ over each of a suitably chosen sequence of contours C_n. Let C_n be a square with center at the origin, sides parallel to the x- and y-axes, and passing through the point $(n + \frac{1}{2})$. Then the points $0, \pm 1, \pm 2, \ldots, \pm n$ are poles of $\cot \pi z$ which lie inside C_n. For any one of these, $z = m$, we have

$$\lim_{z \to m} (z - m) \pi \cot \pi z \, f(z) = f(m) = \text{Res}(m). \tag{84}$$

Let $f(z)$ have simple poles a_k and let the residue of $f(z)$ at a_k be b_k. We assume that no a_k is at any m, and that n is so large that all the

a_k lie inside C_n. Then the residue of $\pi \cot \pi z\, f(z)$ at a_k is $\pi b_k \cot \pi a_k$. Hence by the residue theorem, we have

$$\int_{C_n} \pi \cot \pi z\, f(z)\, dz = 2\pi i \left[\sum_{m=-n}^{n} f(m) + \Sigma\, \pi b_k \cot \pi a_k \right]. \quad (85)$$

To see that $\cot \pi z$ is bounded on C_n, we note that for $y \geq 1$,

$$\cot \pi z = \left| i\, \frac{e^{2ix} + 1}{e^{2ix} - 1} \right| \leq \frac{1 + e^{-2y}}{1 - e^{-2y}} \leq \frac{2}{1 - e^{-2}}. \quad (86)$$

This same bound holds for $y \leq -1$. For the part of C_n with $|y| \leq 1$, we note that $\cot \pi z$ has the period unity, so that we need merely consider $|\cot \pi(\tfrac{1}{2} + iy)|$ for y in this range. We have

$$\left| \cot \pi(\tfrac{1}{2} + iy) \right| = |-\tan iy| = \left| \frac{-i \sinh y}{\cosh y} \right| \leq 1. \quad (87)$$

Next assume that the degree of the denominator of $f(z)$ is at least two more than the degree of the numerator. Then $z^2 f(z)$ will approach a finite limit as $z \to \infty$, so that for n sufficiently large we shall have $|f(z)| < q/n^2$ if z is on C_n. Let M denote the bound for $\cot \pi z$ on all the C_n which we found above. And note that the length of C_n is $8(n + \tfrac{1}{2}) = 8n + 4$. Then by Cauchy's inequality, we find that

$$\left| \int_{C_n} \pi \cot \pi z\, f(z)\, dz \right| \leq \pi M \left(\frac{q}{n^2} \right)(8n + 4) < \frac{9\pi M q}{n}, \quad (88)$$

for $n > 4$. Since the right member tends to 0 as $n \to \infty$, the same is true of the left member, and also that of Eq. (85). Hence the right member of Eq. (85) tends to zero as $n \to \infty$, so that

$$\sum_{m=-\infty}^{\infty} f(m) = -\Sigma\, \pi b_k \cot \pi a_k. \quad (89)$$

The relation $|f(n)| < q'/n^2$ guarantees the absolute convergence of the series on the left, so that any order which eventually includes every term may be used.

If we use $\csc \pi z$ in place of $\cot \pi z$, we obtain in place of Eq. (89) the relation

$$\sum_{m=-\infty}^{\infty} (-1)^m f(m) = -\Sigma\, \pi b_k \csc \pi a_k. \quad (90)$$

If $f(z)$ has multiple poles, or poles a_K which are points m, the only change needed is a direct calculation of the residues, and the omission from the sum of the term $f(a_K)$ for each such a_K.

Example 1. Evaluate
$$S_1 = \sum_{n=1}^{\infty} \frac{1}{n^2}.$$

Solution: Let $f(z) = 1/z^2$. Then the only pole of $f(z)$ is at $z = 0$. Since
$$\cot \pi z = \frac{\cos \pi z}{\sin \pi z} = \frac{1 - \pi^2 z^2/2 + \cdots}{\pi z - \pi^3 z^3/6 + \cdots}$$
$$= \frac{1}{\pi z} - \frac{\pi z}{3} + \cdots, \qquad (91)$$

for $\pi \cot \pi z\, f(z) = (\pi \cot \pi z)/z^2$, Res $(0) = -\pi^2/3$. Here in place of Eq. (89) we have
$$2S_1 = \sum_{m=-\infty}^{\infty}{}' \frac{1}{m^2} = -\text{Res}\,(0) = \frac{\pi^2}{3}. \qquad (92)$$

Hence $S_1 = \pi^2/6$, in agreement with Problem 11 of Exercise 39.

Example 2. Evaluate
$$S_2 = \sum_{n=1}^{\infty} (-1)^{n+1} \frac{1}{n^2}.$$

Solution: Let $f(z) = 1/z^2$. Then the only pole of $f(z)$ is at $z = 0$. Since
$$\csc \pi z = \frac{1}{\sin \pi z} = \frac{1}{\pi z - \pi^3 z^3/6 + \cdots}$$
$$= \frac{1}{\pi z} + \frac{\pi z}{6} + \cdots, \qquad (93)$$

for $\pi \csc \pi z\, f(z) = (\pi \csc \pi z)/z^2$, then Res $(0) = \pi^2/6$. Here in place of Eq. (90) we have
$$-2S_2 = \sum_{m=-\infty}^{\infty}{}' (-1)^m \frac{1}{m^2} = -\text{Res}\,(0) = -\frac{\pi^2}{6}. \qquad (94)$$

Hence $S_2 = \pi^2/12$. We note that from Examples 1 and 2, $\tfrac{1}{2}(S_1 + S_2) = \pi^2/8$, in agreement with Problem 12 of Exercise 39.

Exercise 40

Verify that for $a > 0$, and a not an integer,

1. $\sum_{n=1}^{\infty} \dfrac{1}{n^2 + a^2} = -\dfrac{1}{2a^2} + \dfrac{\pi}{2a} \coth \pi a.$

2. $\sum_{n=1}^{\infty} \dfrac{(-1)^{n+1}}{n^2 + a^2} = \dfrac{1}{2a^2} - \dfrac{\pi}{2a \sinh \pi a}.$

3. $\sum_{n=1}^{\infty} \dfrac{1}{n^2 - a^2} = \dfrac{1}{2a^2} - \dfrac{\pi}{2a} \cot \pi a.$

4. $\sum_{n=1}^{\infty} \dfrac{(-1)^{n+1}}{n^2 - a^2} = -\dfrac{1}{2a^2} + \dfrac{\pi}{2a \sin \pi a}.$

5. $\sum_{n=-\infty}^{\infty} \dfrac{1}{(n + a)^2} = \dfrac{\pi^2}{\sin^2 \pi a}.$

6. $\sum_{n=-\infty}^{\infty} \dfrac{1}{n^4 - a^4} = -\dfrac{\pi}{2a^3} (\cot \pi a + \coth \pi a).$

7. $\sum_{n=-\infty}^{\infty} \dfrac{(-1)^n}{n^4 - a^4} = -\dfrac{\pi}{2a^3} \left(\dfrac{1}{\sin \pi a} + \dfrac{1}{\sinh \pi a} \right).$

8. $\sum_{n=-\infty}^{\infty} \dfrac{1}{n^4 + 4a^4} = \dfrac{\pi}{2a^3} \dfrac{\sinh 2\pi a + \sin 2\pi a}{\cosh 2\pi a - \cos 2\pi a}.$

9. $\sum_{n=-\infty}^{\infty} \dfrac{(-1)^n}{n^4 + 4a^4} = \dfrac{\pi}{2a^3} \dfrac{\sin \pi a \cosh \pi a + \cos \pi a \sinh \pi a}{\cosh 2\pi a - \cos 2\pi a}.$

For r an odd integer,

$$\sum_{n=0}^{\infty} \dfrac{(-1)^n}{(2n + 1)^r} = \tfrac{1}{2} \sum_{n=-\infty}^{\infty} \dfrac{(-1)^n}{(2n + 1)^r}.$$

Use this and Eq. (90) or its modification to deduce that:

10. $1 - \tfrac{1}{3} + \tfrac{1}{5} - \ldots = \pi/4$. This agrees with the result of putting $x = 1$ in the Maclaurin's series for $\tan^{-1} x$.

11. $1 - \dfrac{1}{3^3} + \dfrac{1}{5^3} - \ldots = \dfrac{\pi^3}{32}.$

12. $1 - \dfrac{1}{3^5} + \dfrac{1}{5^5} - \ldots = \dfrac{5\pi^5}{1536}.$

BIBLIOGRAPHY

Ahlfors, L. V.: *Complex Analysis*, McGraw-Hill, New York, 1953.

Courant, R.: *Dirichlet's Principle, Conformal Mapping, and Minimal Surfaces*, Interscience, New York, 1950.

Franklin, P.: *A Treatise on Advanced Calculus*, Wiley, New York, 1940.

Hurwitz, A. and Courant, R.: *Vorlesungen über allgemeine Funktionentheorie und elliptische Funktionen*, Springer, Berlin, 1929.

Knopp, K.: *Theory of Functions* (2 vols.), translated by F. Bagemihl, Dover, New York, 1945.

Kober, H.: *Dictionary of Conformal Representations*, Dover, New York, 1952.

Nehari, Z.: *Conformal Mapping*, McGraw-Hill, New York, 1952.

Springer, G.: *Introduction to Riemann Surfaces*, Addison-Wesley, Reading, Mass., 1957.

Thron, W. J.: *The Theory of Functions of a Complex Variable*, Wiley, New York, 1953.

Titchmarsh, E. C.: *The Theory of Functions*, Oxford, New York, 1939.

Whittaker, E. T. and Watson, G. N.: *A Course of Modern Analysis*, Cambridge, London, 1953.

Recommended Further Reading For 2018 Edition

The number of introductory complex analysis textbooks that currently exists is so large, one is tempted to say it is uncountable. Indeed, with the possible exception of point-set topology, there is no subject of post-calculus higher mathematics that has a larger representation in the textbook literature. This makes recommending further reading in the subject quite difficult because such a bibliography can easily grow to the length of this text itself or beyond! Therefore, to limit this bibliography, I used 3 criteria:

a) Standard texts-that is, textbooks that in my experience, are and have been more widely used at most mathematics programs to teach either undergraduate or graduate complex analysis then other texts;

b) My personal favorites: These are the books that I've used either as a student or a tutor in either learning or teaching complex variables. Author/Editor's prerogative here.

c) Cost: Remember, Blue Collar Scholar's primary goal is to make quality textbooks available inexpensively. In the case of complex analysis, there's a surprisingly large number of cheap textbooks available and I definitely want to include a few of these here.

Using these 3 criteria, I think I came up with a fairly good and useful list of limited size. Since the Franklin text sort of straddles the fence between upper level undergraduate and first year graduate level texts-as many texts like this do-I will include both level texts.

The standard undergraduate textbooks on the subject, as I referenced in the Preface, are Brown/ Churchill (5) and Snider/Saff (6). I must confess to being far more familiar with the former then the latter. In fact, one of the earlier editions-the 5th, I believe-was the textbook for *my* complex variables course. Another confession, though-we didn't really use it much except to do the exercises. Dr. Kulkarni's lecture notes were, in my opinion, far superior for the beginner. But Ravi liked the book quite a bit and encouraged us to read it in addition. What really strikes me now looking back on the text in retrospect-which is not dramatically different the current 8th edition-is how similar the book is to Franklin's in content if not style. The topic selection is virtually identical. The main differences are a)Brown/Churchill is more analytic and less geometric and it's applications are accordingly geared towards differential equations, b) Franklin is a bit more sophisticated and less detailed, many more results are shunted to the exercises. This is not surprising since Franklin's book was aimed at a more sophisticated audience. Still, Brown and Churchill is quite readable and it complements Franklin very nicely. Best of all, the book has not changed dramatically since the first edition was written by Churchill alone in 1941. Brown's *Complex Variables* appeared in 1941 as the one of the first modern complex variable texts for undergraduates and physical science majors. The topics have been standardized: complex numbers, differentiation, analytic and harmonic functions, contour integrals in the plane and physical applications.This means any edition will do and will be far cheaper than a new copy.

The standard graduate text on complex variables, which has been canonized at most top mathematics graduate programs in America, is Lars Alfhors' text (7). The fact this book has achieved such lofty academic status among function theorists baffles me far more then what people see in Ariana Grande. You might think I loathe the book from that initial sardonic remark-but you'd be wrong. I just question its value as an introduction to the subject. The book is dense, dry and quite difficult. Its' prerequisites are much higher than Franklin's-to be able to read, *actually read and learn from*, Alfhors, a student really needs a full year course in real analysis on metric spaces and some knowledge of modern algebra and topology. That being said, the book does have many good qualities. Alfhors is clearly a master and he covers the terrain of complex variables like one. This book emphasizes geometry over analysis in the presentation, but it is far more sophisticated than any of the previously mentioned books. Many concepts from modern topology, abstract algebra and real analysis are used freely and without warning, including brief introductions to homology, germs and sheaves. There are many beautiful but very difficult exercises. Unlike some other textbooks, I can't in all good conscience recommend the earlier editions for a lower price-because *frankly, the first 1953 edition is a train wreck.* The third edition is far superior in both organization and readability. I warn the reader to find an inexpensive international edition to ensure you get the 3rd edition. To be perfectly honest, I think Alfhors deserves its classic status, but I would use it as a second course in complex analysis. I believe the book would be a buzz kill for all but the very strongest graduate students without some prior knowledge of the subject. In fact, I think Franklin could serve very well as preliminary preparation for tackling Alfhors.

Believe it or not, the biggest problem a lot of people have these days using Alfhors as a graduate text is that it's now out of date. The most up-to-date, intensive and yet readable graduate textbook I know on the subject is Narasimhan and Nievergelt's text (8). It is complex analysis for the *serious* graduate student planning to do research in analysis. The first edition of the book developed from Narasimhan's infamous first year course at the University of Chicago that was boot camp even for the graduate students at the U of C-some of the strongest mathematics students in the world. The second edition added a large and excellent collection of exercises that Narasimhan's former student Nievergelt developed for the original notes when using them as the text for his own graduate course at the University of Michigan. I think it would be academic suicide to try and learn complex variables from scratch from this book-and it's not intended to. It's clearly intended for students with some prior knowledge of the subject. From a rapid review of the basics of analytic functions, covering spaces and Runge's theorem through the basics of functions of several complex variables and the elements of complex manifolds. There are a number of references to both the textbook and research literature at the end of each chapter, vastly expanding the actual coverage of the text to the frontiers of the subject. (Apparently this is still true despite the book now being nearly 20 years old!) Unlike "graduate" texts like Alfhors, which can indeed be studied by very strong undergraduates, Narasimhan and Nievergelt is unquestionably a graduate level text even for well-prepared students. It requires not only some basic knowledge of complex analysis, but a good working knowledge of real analysis on metric spaces at the level of Rudin's *Principles of Mathematical Analysis* as well as a basic knowledge of abstract and linear algebra and topology. Later chapters require some knowledge of the theory of integration and functional analysis. Indeed, many important basic results are in the exercises. In short, it's a **serious** book for advanced students by the University of Chicago master and his former student. A strong student about to take first year graduate complex analysis out of this book would do themselves a great service studying Franklin or another of the recommended undergraduate texts here in the summer before.

You can't seriously attempt to write a recommended reading list of textbooks on complex analysis without at least mentioning the amazing treatise on the subject written by the Russian master, A. I. Markushevich's and translated by Richard Silverman into English (9). Markushevich was one of the world's eminent teachers and researchers in analytic functions at the height of the Cold War and the pinnacle of the Golden Age of Soviet mathematics and physics centered at Moscow State University's Department of Mathematics And Mechanics. Silverman for many years was one of the foremost translators of Russian mathematics textbooks in America. This epic textbook on classical complex analysis was based on Markushevich's decades of teaching complex analysis to both undergraduates and graduate students there. I stumbled upon this classic in the old dusty library of Queens College, where I spent many days and nights attempting to master both mathematics and my own life-and fell in love with it immediately. The sheer range of the book is staggering: the arithmetic and basic geometry of complex numbers, sequences, limits and derivatives, the elementary geometry of the Argand plane, derivatives, contour integrals, infinite series and convergence, Laurent series and the calculus of residues with many applications, univalent and harmonic functions, applications to physics such as fluid dynamics, entire and meremorphic functions, Riemann surfaces and analytic continuation *and much, much more*. The prerequisites for the book are about the same as Franklin and like Franklin, the authors strive for a balance between the analytic and geometric aspects. In its' original English edition published by Prentice-Hall, the book was organized into 3 volumes. In the 2nd edition published by the American Mathematical Society, it's now a single massive volume with a comprehensive table of contents and index.

While I don't see much difference in content between both, the 2nd edition would clearly be easier for a student to use. In my opinion, one of the most complete and readable works in the textbook literature on complex variables-indeed, on any subject. The only ones who might have a problem with it are graduate students who need more modern topics in the subject such as complex manifolds and Julia sets, which clearly this book won't have. But that material can easily be found in other sources. It is one of my favorite textbooks period and it makes a fantastic reference as well. But sadly, it's very expensive, which is why I don't give my unequivocal endorsement here. If you can afford a copy, by all means do so. At the very least, you should do what I did and borrow it.

I promised making quality mathematics textbooks available inexpensively is the main mission statement of BCS books. As a result, I try very hard also to recommend inexpensive books as supplementary study whenever I can in our books. This book by Franklin is no different. There's a surprisingly large number of such books available for complex variables, most published by our venerable "predecessor" company, Dover.

2 very inexpensive texts that each can complement Franklin very effectively or be used to supplement one of the more sophisticated texts here are Flanagan (10) or Fisher (11). Flanagan's and Fisher's books are essentially Brown/Churchill clones, as many beginning texts pitched at this level are. For the cash-strapped student for whom this bibliography is intended, this is very good thing indeed because it means there are a lot of comparable low-cost alternatives to Brown/Churchill out there. Both Fisher and Flanagan are good ones and they're quite different in style from each other despite covering much the same ground. Flanagan first. This is a book that emphasizes the connection between complex analysis and calculus in the real plane. As a result, its prerequisites are about the same as Franklin: calculus up to the basics of multivariable calculus and some rigorous analysis. However, as the subtitle implies, it takes a more analytic perspective then Franklin, which strives hard to balance both aspects. The fact that $\mathbf{R}^2$ is in one to one correspondence with $\mathbf{C}$ and the fact complex analytic functions satisfy harmonic conditions are both used extensively throughout the book to establish the properties of complex functions. The emphasis on calculus has the advantage that it enables the author to explain many subtle points of complex analysis that might be lost on a beginner- such as the difference between complex-analytic and real-analytic functions, the integration of meromorphic functions and applications of complex variables to differential equations. Fisher's book has about the same prerequisites, but it's a bit less sophisticated and thorough. Explanations and derivations are more detailed. It also has significantly more examples, especially in the second half of the book. It also emphasizes the algebraic and plane geometric aspects of the subject a bit more then Flanagan. Contour integration is not discussed in as great a detail as Flanagan, but analytic functions as conformal mappings of the plane are discussed at considerably more length. Fisher also has a larger and more diverse body of exercises. Judging them both, Fisher is better for the less prepared student, but

I think Flanagan would probably prepare a student better for further study. I think they complement each other very nicely. If I could only afford one, I'd take Flanagan over Fisher. But since both are so cheap, why wouldn't you have both?

I'm now moving on to a rather oddball pair of books available from Richard Silverman ((12) and (13)) I'm listing both these books under the same entry since in my opinion, you can't get the full benefit for a course in complex variables without having both of them. There are actually 2 versions of this book published by Dover-the former text is a book emphasizing the purely mathematical aspects of basic complex analysis and the latter is similar but somewhat different book that emphasizes the physical applications such as to fluid mechanics. In my opinion, (12) is a highly underrated book and really deserves a closer look. Most people, like me, look at it and knowing Silverman's history, naturally-but incorrectly-assume either book is simply a watered down or "Reader's Digest" version of Silverman's translation of Markushevich's text. This is very unfortunate since it's not true-these are both original textbooks on complex analysis. They are both clearly heavily influenced by the Russian master's classic, but still very much Silverman's own work. As expected, since both are introductory books on complex variables, there's considerable overlap between them. But they diverge as the former focuses on topics in classical function theory traditionally more of interest to pure mathematicians-such as residues, singularities and meremorphic functions-and the latter focuses on applications of interest to the physicist and/or engineer, such as harmonic functions, conformal mapping and physical applications like fluid dynamics. To be honest, the second book seems more polished and comprehensive, but the first takes a more rigorous presentation. Hence, to really benefit from both books, it really pays to get them both. And again, since they're so cheap, why not?

We now come to the 2nd edition of the book by Ash and Novinger (14) This is a dramatic revision of a graduate textbook that Ash and Novinger wrote in the 1970's and it contains a concise but thorough graduate-level course in complex variables. It contains all the material you'd expect in this course: the topology of the complex plane, analytic functions, convergence of complex power series, harmonic functions, the Cauchy theorem, etc. Again, it's done at a graduate level sophistication, so it assumes knowledge of an intermediate real variables course on abstract metric spaces. There are no physical applications, but very clear discussions and lots of exercises with proofs. It's a bit dry, but that's to be expected to some degree in a pure mathematics course at this level. It's a good book, but it'll never replace the great classics.

I'd be derelict in my duty if I didn't recommend this famous collection of solved exercises in complex analysis (15). It's hard to believe, but up until recently this classic problem book that was the standard accompaniment to analytic function theory courses in the old Soviet Union and whose translation was eagerly awaited in this country-went out of print and copies could be found on the internet in the late 1990's for 200-300 dollars. Fortunately, this was a brief period and Dover has made this classic available cheaply again. Why this book was so sought after isn't hard to see-it has an enormous wealth of very diverse problems complete with solutions, ranging from basic analytic function theory to advanced topics like analytic continuation and Riemann surface theory to physical applications to fluid flow and elasticity. Indeed-many of the exercises in the English translation of Markushevich originated in the Russian original of this book! Any good math major or graduate student who doesn't buy a copy and work through it before their qualifying exam in complex variables is an idiot. Sorry, at this price, you're an idiot if you don't get one!

I've saved the best for last. As far as I'm concerned, there's no better one-volume introduction to the beautiful subject of complex analysis then Gamelin's book (16). It's actually pretty stunning what Gamelin has achieved in this book. Requiring only a good background in multivariable calculus-and not even rigorous calculus at that!-he builds virtually the entire subject of complex variables of one variable and develops all the needed tools from advanced mathematics as he goes. The book developed from the author's various lecture notes for myriad courses in the subject over 40 years, primarily at UCLA but other places where he was a visiting lecturer. His primary problem with teaching complex analysis over the years has been the students usually have very diverse backgrounds, ranging from average talent post-calculus undergraduates in math and physics to very strong PhD students in analysis. His solution to this corundum was to present the subject with minimal prerequisites, but to present sophisticated topics several times at different levels of difficulty. He particularly uses this for topics he judges absolutely critical for an understanding of complex variables, such as Riemann surfaces and analyticity. The book is organized into 3 sections. The first "unit" is intended for a basic undergraduate course, requiring only calculus. It covers complex numbers, elementary functions, the topology of the complex plane, analytic functions, contour integration and harmonic functions. The second and third parts build slowly for strong undergraduates, beginning graduate students and finally to PhD exam level topics such as Julia sets and meromorphic functions. It also has one of the most complete and basic presentations of Riemann surfaces you'll ever find in a complex analysis text. It's beautifully written with a host of examples, pictures and a ton of excellent problems. It also has many applications to geometry and physics. There are a legion of great books on complex analysis at various levels and one of my favorites is the one by Franklin you now hold- that's why I went through the trouble of republishing it. But if I had to learn complex analysis by a fixed deadline, was forced to choose only one book and had to make sure I learned it right, *this* is the one I'd pick.

References:

1) http://www-history.mcs.st-andrews.ac.uk/Biographies/Franklin.html

2) Rota, Gian-Carlo; Palombi, Fabrizio, *Indiscrete Thoughts*, Birkhäuser, 2008

3) Fite, William Benjamin; Maestro, Karo, *Old School Advanced Calculus*, Blue Collar Scholar/ Createspace, 2018

4) Miller, Kenneth S.; Maestro, Karo, *Vector Analysis: A Supplement To Old School Advanced Calculus*, Blue Collar Scholar/ Createspace, 2018

5) Brown, James Ward; Churchill, Ruel W., *Complex Variables And Applications*, 9th ed, McGraw-Hill, 2013

6) Saff, Edward; Snider, Arthur D., *Fundamentals of Complex Analysis: with Applications to Engineering and Science,* 3rd edition, Pearson, 2017

7) Alfhors, Lars, *Complex Analysis*, 3rd ed, McGraw-Hill, 1979

8) Narasimhan, Raghavan, Nievergelt, Yves, *Complex Analysis Of One Variable*, 2nd ed, Birkhäuser, 2000

9) Markushevich, A.I., *Theory of Functions of a Complex Variable,* edited and translated by Richard Silverman, 2nd ed, American Mathematical Society, 2005

10) Flanagan, Francis j., *Complex Variables: Harmonic and Analytic Functions* Dover Publications, 1983

11) Fisher, Stephen D., *Complex Variables,* 2nd ed, Dover Publications, 1999

12) Silverman, Richard, *Introductory Complex Analysis*, Dover Publications,1984

13) Silverman, Richard, *Complex Analysis With Applications*, Dover Publications, 2010

14) Ash, Robert, Novinger, W.P., *Complex Variables*, 2nd ed, Dover Publications, 2007

15) Volkovyskii ,L. I., Lunts,G. I, Aramanovich, I. G., *A Collection of Problems on Complex Analysis,* 2nd Rev. Ed., Dover Publications, 2011

16) Gamelin, Theodore W., *Complex Analysis*, Springer-Verlag, 2001

INDEX

Numbers in parentheses refer to problems, the first of which begins on the page whose number immediately precedes the parentheses. Other numbers refer to pages.

A

Absolute convergence, 37
   of power series, 43
Absolute value, 2, 8
Addition theorems, 171
   for the exponential function, 51
   for hyperbolic functions, 62
   for trigonometric functions, 56, 175 (1)
Argument, 10
Argument principle, 198
Analytic continuation, 170
   definition by, 172
   using reflection, 177 (12)
Analytic function, 24, 160
Associative law, 3

B

Bernoulli numbers, 157 (14-16)
Bessel's functions, 158 (18, 19)
Bessel's inequality, 163
Beta function, 230 (7)
Bilinear transformations, 98, 100
   group of, 105 (4)
   particular, 116, 119 (8-12)
Binomial expansion, 48 (10), 74 (2), 136 (8), 155 (7, 9)
Boundary, natural, 177 (11)
Boundary point, 17
Boundary value problems, 91, 95 (7-16)
Bounded set, 17
Branch cut, 71, 172
Branch point, 72, 172
   at infinity, 73, 172

C

Cauchy-Goursat integral theorem, 133, 136, 139 (9-11), 142
Cauchy-Riemann equations, 26
   in polar form, 29 (5)
Cauchy's inequalities, 162
Cauchy's integral formula, 140
   for the nth derivative, 152
Cauchy's integral theorem, 132
Cauchy's principal value, 221
Cauchy's residue theorem, 196, 210
Cauchy's rule for convergence, 38
Cauchy's rule for products, 47 (5)
"Circle," 102
Circle of convergence, 38
Cis, 10
Closed set, 16
Closure, 17
Coaxial circles, 112
Commutative law, 3
Complex numbers, 1, 2, 8
Complex plane, 8
Complex variable, 20

INDEX

Conformal transformations, 77, 83
  of domains, 205
Conjugate complex numbers, 5
Conjugate harmonic functions, 31
Connected set, 17
Continuity, 21
  of sum of series, 42
Contours, 127
  indented, 222
  special types of, 228
Convergent series, 36
  absolutely, 37
  uniformly, 41
Covering set, 18
Critical point, 87
Cross ratio, 114 (9, 10)
Cut, branch, 71, 172

D

De Moivre's theorem, 13
Derivative, 23
Differentiable function, 21
Differentiation of power series, 43
Differentiation rules, 24
Dirichlet's problem, 92
Distributive law, 3
Divergent series, 36
Domain, 17
Duhamel's principle, 126, 128 (1)

E

Elementary functions, 50, 70
Entire functions, 51, 165, 174
Essential singularity, 188, 210
  nonisolated, 191, 195 (9-11)
Euler's relations, 52, 56
Even functions, 56, 62
Exponential functions, 50

F

Fresnel's integrals, 231 (11)
Functions, 20
Fundamental region, 53, 60 (12, 26), 63 (9, 13)

Fundamental theorem of algebra, 166, 169 (20, 21), 202 (13), 207 (1)

G

Gamma functions, 230 (18)
Green's theorem, 132

H

Harmonic functions, 31, 91
Heine-Borel theorem, 18
Holomorphic functions, 24
Hyperbolic functions, 57, 61

I

Imaginary component, 2
Imaginary numbers, pure, 2, 4, 8
Imaginary unit, 1, 4
Implicit function theorem, 80
Indented contours, 222
Inequality for absolute values, 8 (13, 14)
Inequality for integrals, 130
Inequality for series coefficients, 162
Infinite series, 35
Infinity, point at, 103
  branch, 73, 172
Integrals, 121
  complex along curved paths, 124
  of derivatives, 129
  evaluated by residues, 209, 213, 215, 218, 226
  inequality for, 130
  linear properties of, 127
  principal value for, 221
  real definite, 122
  theorem of Cauchy on, 132
  theorem of Cauchy-Goursat on, 133, 136, 139 (9-11), 142
  of uniformly convergent series, 140
Interior point, 16
Inverse functions, 65, 69, 84
  hyperbolic, 70, 74 (11-13)
  trigonometric, 70, 74 (8-10, 18, 19)

INDEX

Inversion, 101
   in any circle, 108
Isogonal transformation, 87

J

Jacobian determinant, 79

L

Laplace's equation, 30, 91
   in polar form, 33 (17)
Laurent's expansion, 179, 180
   uniqueness of, 183
Legendre's polynomials, 158 (17)
Limit point, 16
Linear transformations, 89
Liouville's theorem, 165
   generalized, 167
   for harmonic functions, 169 (19)
Logarithmic functions, 64
   expansions of, 156 (10), 175 (2-5)

M

Maclaurin series, 46, 152
Magnification, 86, 99
Maximum modulus principle, 164
Milne-Thompson, method of, 33 (23)
Minimum modulus principle, 165
Monodromy theorem, 173
Morera's theorem, 159
Multiply connected regions, 139

N

Natural boundary, 177 (11)
Neighborhood, 16

O

Odd functions, 56, 62
Open set, 17
Ordered pairs, 2
   operations on, 3

P

Parseval's identity, 163
Partial fraction expansion, 196 (14)
Path of integration, 123
Periodic functions, 52, 57, 61 (26), 63 (5)
Permanence of form, 171
Point sets, 16
Poisson's integral, 145 (13), 168 (12)
   for a half-plane, 146 (14)
Polar form, 9
Pole, 187, 210
   of order m, 187, 210
Polynomials, 25, 168, 192
Power series, 35, 37
   as Maclaurin's or Taylor's, 46, 152
   uniqueness of, 46, 48 (12, 17), 152, 157 (13)
Powers of complex numbers, 13, 68
Principal branch of logarithm, 64
   Log z, 67 (5, 6)
   Log $(1 + z)$, 156 (10)
Principal part, 187
   at infinity, 190
Proper parameter, 123
Product expansions, 235 (9, 10)

R

Radius of convergence, 38, 152
Rational fractions, expansions in, 196 (14), 232, 235 (1-3, 7, 8)
Rational functions, 25, 192
   degree for, 204
   residues of, 203
Real component, 2
Real numbers, 2, 8
Reflection, 110
   analytic continuation by, 177 (12)
Region, 17
   closed, 17
   multiply connected, 139
   open, 17
   simply connected, 138
Regular analytic function, 24
Remainder after N terms, 40

INDEX

Residues, 187, 188, 209
  applications of, 209
  Cauchy's theorem on, 196, 210
  evaluation of, 194 (3–5), 211
  at infinity, 190, 203, 210
  of rational functions, 203
Riemann surfaces, 71
Riemann's theorem, 188
Rotation and magnification, 86, 90
Rouché's theorem, 200

S

Schwarz-Christoffel transformation, 206, 208 (4–6)
Simple curve, 129 (11)
Simply connected region, 136
Single-valued function, 21
Singular points, 25, 174, 186
  on circle of convergence, 174
  isolated, 188, 209
  removable, 187
Stereographic projection, 107 (18–23)
Summation of series, 236

T

Taylor's expansion, 46, 148, 149

Transformations, 78
  approximating, 79, 88
  bilinear, 98, 100
  conformal, 77, 83, 205
Trigonometric functions, 55
  product for $\sin z$, $\cos z$, 235 (9, 10)
  series for $\tan z$, $z \cot z$, 158 (15, 16)

U

Uniform continuity, 22
Uniformly convergent series, 40
  with continuous sum function, 42
  integration of, 149

V

Variable, complex, 20
Vector, 9

W

Weierstrass's M-test, 41
Weierstrass's theorem, 189, 194 (6, 7)

Z

Zero, 58
  of order m, 188

About the Author

Phillip Franklin was one of the most famous American-born mathematicians of the first half of the 20th century. He was one of the first American mathematicians with a truly modern training; getting his doctorate at Princeton University under Oswald Velben in 1922 and beginning his career by spreading knowledge of the frontiers of mathematics to the other top programs. Marshall Stone famously commented that the entire Harvard University department-students and faculty alike- learned the elements of topology from the seminar Franklin taught when he arrived there as a new PhD. He then moved on to the Massachusetts Institute of Technology where we spent the bulk of his career, both teaching and doing research. Franklin became well known as an analyst, but the truth is he published in a number of fields, including geometry, number theory, graph theory and topology. He's best known for authoring some of the most well-received and influential textbooks of his generation. Among his most famous texts were *Differential equations for electrical engineers* (1933), the exhaustive undergraduate analysis text *A Treatise on advanced calculus* (1940), which quickly became one of the classic standard texts for advanced calculus published before Walter Rudin's *Principles of Mathematical Analysis*, its companion sequel for physics and engineering students*, Methods of advanced calculus* (1944), the standard introductory calculus text *Differential and integral calculus* (1953) and the honors calculus text *Compact calculus* (1963). Several of these books are now being republished, which the publisher considers a very good thing indeed that he's attempting to be part of.

About the Publisher/ Editor:

Karo Maestro and The Mathemagician are both *nome-de-plume* for a former graduate student in mathematics. His true identity will remain a secret for now, but one day will be revealed to all. Some things this enigma can reveal: He was a distinguished undergraduate student as a double major in mathematics and biochemistry whose poor health and personal tragedies prevented completing graduate studies. Unbowed and undaunted, he plans to return to ultimately obtain a PhD in pure mathematics before dying. Partially to that end, he is building Blue Collar Scholar, a publishing company committed to making high quality sources of mathematics-both original works and reprints-available widely and inexpensively to students of all backgrounds. The volume you hold is the fifth published with more to follow. He has 3 bachelors' degrees, in philosophy, physical chemistry and mathematics as well as minors in biochemistry and psychology. He is also a past reviewer of textbooks for the Mathematical Association Of America. Among his more recent achievements are the website, TULOOMATH (www.tuloomath.com), which is designed to be a one stop hub for free downloadable lecture notes and online textbooks in university mathematics from high school algebra to PhD level topics. He's also the author of *Tables,Chairs And Beermugs* , the associated blog for the website where he reviews textbooks and vents on matters mathematical and academic. He's painfully blunt, opinionated and has a comment on just about everything in creation. He's painfully blunt, opinionated and has a comment on just about everything in creation-from math textbooks to progressive politics to how to make a perfect cup of organic green tea. He loves JRR Tolkien, science fiction, comic books and movies of all kinds, the comedy of the late great George Carlin, playing with his wonderful nieces, real barbeque, creative burgers and fresh cut French fries, tall curvaceous women and popular music, particularly Bon Jovi and U2. Among his current musical favorites are Alysia Cara, Adele, Imagine Dragons and Ed Sheeran. He is currently working on 2 original books on how to earn a PhD in pure mathematics by self-study alone.

(No, he's not insane-at least he doesn't think so.)

And he still has no clue what people see in Ariana Grande. Not a clue.

www.ingramcontent.com/pod-product-compliance
Lightning Source LLC
Chambersburg PA
CBHW062212220526
45471CB00009B/3172